Raspberry Pi and MQTT Essentials

A complete guide to helping you build innovative full-scale
prototype projects using Raspberry Pi and MQTT protocol

Dhairya Parikh

‹packt›

BIRMINGHAM—MUMBAI

Raspberry Pi and MQTT Essentials

Group Product Manager: Rahul Nair
Publishing Product Manager: Meeta Rajani
Senior Content Development Editor: Sayali Pingale
Technical Editor: Rajat Sharma
Copy Editor: Safis Editing
Project Manager: Neil Dmello
Proofreader: Safis Editing
Indexer: Subalakshmi Govindhan
Production Designer: Shyam Sundar Korumilli
Marketing Coordinator: Nimisha Dua
Senior Marketing Coordinator: Sanjana Gupta

First published: August 2022

Production reference: 0240822

Published by Packt Publishing Ltd.
Livery Place
35 Livery Street
Birmingham
B3 2PB, UK.

978-1-80324-448-8

www.packt.com

To my parents, Bhumika and Atit, for their sacrifices and constant encouragement.

Contributors

About the author

Dhairya Parikh currently works as a data engineer at Accenture, where he builds efficient data products to help their clients get the most out of their data. He completed his undergraduate studies in electronics engineering at BVM Engineering College, Anand. He is a seasoned project developer with several award-winning projects under his belt. His most recent literary works include several articles written for the Circuit Cellar magazine. Most of his articles and projects are based on IoT and machine learning. He is an open source enthusiast with an interest in building projects that make a positive impact on people's lives using new development hardware and writing about them in his spare time.

I would like to thank my parents and family for their constant support and encouragement throughout this project.

I would also like to thank Neil D'mello, Shazeen Iqbal, Preet Ahuja, Rafiaa Khan, and the entire Packt team for their guidance throughout this project, and Asim Zulfiqar and John Witts for their technical insights! Finally, I would like to thank Yashashree Hardikar for all the initial support.

About the reviewer

Asim Zulfiqar is a blogger and tech content creator who has been writing tutorials on embedded systems and IoT, specifically on MQTT, on his blog and YouTube channel, *High Voltages*. He also provides IoT and embedded freelance services to companies through this channel.

He completed his bachelor's degree in electronic engineering at Sir Syed University of Engineering and Technology, Pakistan. After that, he completed his Erasmus Mundus joint master's degree program in Photonics Integrated Circuits, Sensors, and Networks at Scuola Superiore Sant'Anna (Italy), Aston University (UK), and Osaka University (Japan).

Reviewing a book is more exciting than I thought, and I have enjoyed reviewing this book. None of this would have been possible without my family, friends, teachers, and all the open-source communities that helped me gain enough knowledge to do so. I would also like to thank the author of this book, Dhairya, who helped me learn some new concepts, and Packt, for providing me this opportunity.

Table of Contents

3

Introduction to ESP Development Boards

4

Node-RED on Raspberry Pi

Part 2: Practical Implementation – Building Two Full-Scale Projects

5

Major Project 1: IoT Weather Station

6

Major Project 2: Smart Home Control Relay System

Part 3: How to Take Things Further – What Next?

7

Taking Your MQTT Broker Global

8

Project Prototype to Product – How?

Preface

The future of IoT has the potential to be limitless. By 2025, it is estimated that there will be more than 21 billion IoT devices. So, wouldn't it be great if you could add these to your known technological stacks? But where to start? Of course, with the basics.

First, we will learn about the most popular hardware used for IoT prototyping, the Raspberry Pi. Then, we will learn what MQTT, one of the most used communication protocols for communicating between devices, is. We will then explore why these are the most suitable options to get started, their advantages, and how they are currently being used in the industry. Then, you will see how to use them together by setting up your very own MQTT Server on the Raspberry Pi and understanding how it works. We will get into the details of MQTT and learn more about the clients or devices we will connect to our server. In particular, we will cover two very popular IoT development boards among project developers: ESP8266 and ESP32. You will also learn how to build interactive dashboards on your Pi and control or monitor your client devices. You will build the dashboards using another popular software – Node-RED.

You will then put your theory into practical use by creating two full-scale projects: an IoT weather station and a smart relay system. That's not all; you will also learn how to host your very own MQTT server on a virtual cloud service. Then you will be guided on the next steps to take after reading this book, what technologies to learn along with some useful project recommendations. Finally, we will cover the popular cloud platforms (AWS and GCP) to create IoT projects and also create a project where we connect our Node MCU to AWS IoT.

Who this book is for

This book is suitable for a wide range of audiences. Particularly, this book is targeted at students who want to start building IoT projects, educators who want to teach an introductory IT course, technology enthusiasts, and IoT and hardware developers.

What this book covers

Chapter 1, *Introduction to Raspberry PI and MQTT*, provides an introduction to the hardware we will be using, the Raspberry Pi. Moreover, it will also cover the basics of MQTT and how the communication protocol actually works. Next, we will learn to set up the Raspberry Pi. This includes installing the popular Debian-based Raspberry Pi OS on our Raspberry Pi. After that, we will install the necessary libraries and packages to make our device a local MQTT broker.

Chapter 2, MQTT in Detail, covers how exactly MQTT works. This includes a gentle introduction to MQTT brokers and clients, and different MQTT control packets will be covered in detail. Finally, we will see a demonstration of how a client connects and communicates with a broker.

Chapter 3, Introduction to ESP Development Boards, is all about implementing what we learned in the previous chapter. It will first introduce you to the popular ESP development boards – NodeMCU and ESP32. After covering the specifications of each board, we will move on to learn how to set up the boards as an MQTT client. Finally, we will create our first project wherein we will connect to our Raspberry Pi's MQTT broker and control the onboard LED through MQTT.

Chapter 4, Node-RED on Raspberry Pi, gets you acquainted with very popular software for the Raspberry Pi – Node-RED. It is divided into four sections. First is an introduction to Node-RED, followed by a guide to installing and setting up Node-RED on Raspberry Pi. After that, we will cover the Node-RED MQTT and dashboard components, and then create a simple project to implement everything we have learned.

Chapter 5, Major Project 1: IoT Weather Station, is where, now that we have the knowledge from all the topics discussed in the previous chapters, we will be working on our first major project: making an IoT weather station. This chapter provides step-by-step instructions on how to build this.

Chapter 6, Major Project 2: Smart Home Control Relay System, helps you create a smart home device to control wall switches using the Node-RED dashboard hosted on the Raspberry Pi. The device will be based on the popular ESP32 development board. For this project, we will be preparing a PCB instead of creating the circuit on a breadboard for a more finished and professional look.

Chapter 7, Taking Your MQTT Broker Global, is where we will discuss the advantages of having an online MQTT broker further, and two major options that we currently have to achieve these advantages. We can still use the local broker on our Pi, but we can route all the data to any destination through the internet.

Chapter 8, Project Prototype to Product, How?, starts by exploring IoT services provided by some popular cloud services, such as AWS and GCP, now that the book has covered all the essentials required to get you familiar with all the concepts related to Raspberry Pi and MQTT. We will even create a project demo integrating our Node MCU board with AWS IoT.

To get the most out of this book

This book has been written for beginners, so in terms of knowledge, there are no prerequisites. As for the hardware, you will need all the hardware devices listed in the following table in order to follow along and build projects with me. In terms of software requirements, you will need the Raspberry Pi Imager software (available for all three major operating systems) to create flashed SD cards for your Pi (it even supports SD card formatting), Wireshark on the Raspberry Pi OS to dissect the MQTT control packets, and Node-RED as a dashboard interface for our projects, also to be installed on Raspberry Pi OS.

Hardware covered in the book	Operating system requirements
Raspberry Pi	Windows, macOS, or Linux for your main system
ESP8266-based NodeMCU development board	Raspberry Pi OS for Pi
ESP32 development board	

If you are using the digital version of this book, we advise you to type the code yourself or access the code from the book's GitHub repository (a link is available in the next section). Doing so will help you avoid any potential errors related to the copying and pasting of code.

Download the example code files

You can download the example code files for this book from GitHub at `https://github.com/PacktPublishing/Raspberry-Pi-and-MQTT-Essentials`. If there's an update to the code, it will be updated in the GitHub repository.

We also have other code bundles from our rich catalog of books and videos available at `https://github.com/PacktPublishing/`. Check them out!

Download the color images

We also provide a PDF file that has color images of the screenshots and diagrams used in this book. You can download it here: `https://packt.link/860jg`.

Conventions used

There are a number of text conventions used throughout this book.

`Code in text`: Indicates code words in text, database table names, folder names, filenames, file extensions, pathnames, dummy URLs, user input, and Twitter handles. Here is an example: "Mount the downloaded `WebStorm-10*.dmg` disk image file as another disk in your system."

A block of code is set as follows:

```
void setup ()
{
  pinMode (LED_BUILTIN, OUTPUT);
}
```

Any command-line input or output is written as follows:

```
sudo apt install mosquitto mosquitto-clients
```

Bold: Indicates a new term, an important word, or words that you see onscreen. For instance, words in menus or dialog boxes appear in **bold**. Here is an example: "Select **System info** from the **Administration** panel."

> **Tips or Important Notes**
> Appear like this.

Get in touch

Feedback from our readers is always welcome.

General feedback: If you have questions about any aspect of this book, email us at customercare@packtpub.com and mention the book title in the subject of your message.

Errata: Although we have taken every care to ensure the accuracy of our content, mistakes do happen. If you have found a mistake in this book, we would be grateful if you would report this to us. Please visit www.packtpub.com/support/errata and fill in the form.

Piracy: If you come across any illegal copies of our works in any form on the internet, we would be grateful if you would provide us with the location address or website name. Please contact us at copyright@packt.com with a link to the material.

If you are interested in becoming an author: If there is a topic that you have expertise in and you are interested in either writing or contributing to a book, please visit authors.packtpub.com.

Share Your Thoughts

Once you've read *Raspberry Pi and MQTT Essentials*, we'd love to hear your thoughts! Scan the QR code below to go straight to the Amazon review page for this book and share your feedback.

https://packt.link/r/1803244488

Your review is important to us and the tech community and will help us make sure we're delivering excellent quality content.

Part 1: Covering the Basics

After completing this section, you will know all about Raspberry Pi, MQTT, NodeMCU, ESP32 development boards, and Node-RED. You will gain the knowledge required to build intermediate-complexity projects.

This part comprises of the following chapters:

- *Chapter 1, Introduction to Raspberry PI and MQTT*
- *Chapter 2, MQTT in Detail*
- *Chapter 3, Introduction to ESP Development Boards*
- *Chapter 4, Node-RED on Raspberry Pi*

1
Introduction to Raspberry Pi and MQTT

In recent years, the **Internet of Things (IoT)** has been a trending field for research and development. The future of IoT has the potential to be limitless. By 2025, it is estimated that there will be more than 21 billion IoT devices. So, wouldn't it be great if you could add these to your known technological stacks? In this book, we will start with the absolute basics.

I will walk you through two fascinating subjects throughout this book: **Raspberry Pi**, which is a prevalent development board for beginners, and **MQTT**, a very commonly used and robust communication protocol to delve into the world of IoT.

This chapter will introduce you to the basics of MQTT and Raspberry Pi. Moreover, it will help you set up your Raspberry Pi. Although simple, it is crucial to perform each step as this will help us set up our own **local MQTT broker** on the Raspberry Pi. This will also help you understand how to get started with your new Raspberry Pi by installing an operating system onto it.

First, we will flash the popular Debian-based Raspberry Pi OS on our Raspberry Pi. After that, we will install all the necessary libraries and packages to make our device a local MQTT broker.

We will cover the following main topics in this chapter:

- What is MQTT and how does it work?
- A gentle introduction to Raspberry Pi
- Setting up your Raspberry Pi

So, let's start by knowing what MQTT is.

What is MQTT and how does it work?

In this section, we will learn about the essential concepts of MQTT. First, we will look at the basic concepts of MQTT and some history, followed by the functionality and components of MQTT. Finally, we will have a brief encounter with the salient features of MQTT.

Please note that there are different versions of MQTT, and most of what we discuss is relevant to the MQTT protocol version 3.1.

What is MQTT?

MQTT stands for **Message Queuing Telemetry Transport**. It is a lightweight communication protocol.

According to the official *MQTT v3.1* documentation:

> *MQTT is a Client-Server publish/subscribe messaging transport protocol.*
> *It is lightweight, open, simple, and designed to be easy to implement. These*
> *characteristics make it ideal for use in many situations, including constrained*
> *environments for communication in Machine to Machine (M2M) and Internet*
> *of Things (IoT) contexts where a small code footprint is required, and network*
> *bandwidth is at a premium.*

This is a clear and clean definition of the MQTT protocol in just a few lines. It is a messaging protocol designed for easy implementation, primarily client side. It is an open and lightweight communication protocol with minimal packet overhead. It is generally used for communication between two or more devices.

Basic concepts of MQTT

Now that we know what MQTT is, we will explore the basic concepts of MQTT we came across in the previous section. More specifically, we will look at MQTT as a **publish/subscribe protocol**.

What exactly is a publish/subscribe protocol?

The publish/subscribe protocol is an alternative to the traditional client-server architecture. It means that instead of categorizing both sending and receiving machines as clients, the clients who send a message are publishers and the clients who receive the messages are the subscribers.

Another essential feature of such protocols is the decoupling between the clients. In simple words, the clients never directly communicate with each other. They are mediated by the third component of this system, known as the broker. In this book, we will be using our Raspberry Pi as an MQTT broker, which connects different client devices within a local network. The primary function of the broker is to mediate and manage all communications between the various clients (i.e., *publishers* and *subscribers*).

To better understand how the whole system works, please see *Figure 1.1*, which shows how the communication protocol operates with a very simplified diagram. In this example, the publishing client is a temperature monitor and the subscribing device is a mobile phone:

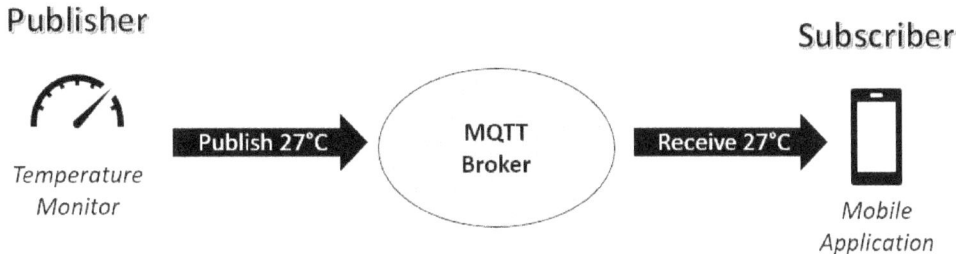

Figure 1.1 – Basic MQTT communication flow

Please note that this is just a simplified representation. There can be multiple publishers and subscribers connected to a single broker. As you can see, the temperature monitor sends the current temperature value of 27°C through the MQTT communication protocol, which is then received by the MQTT broker, which routes it to the subscriber, a mobile application in our case.

We will now look into some details about publishers, subscribers, and brokers with the help of an example:

- **Publishers**: These devices or machines are responsible for sending the collected data to the brokers. For instance, if you have an air quality monitoring system that monitors the CO_2 levels in the air every 30 seconds, the device will be set to publish the CO_2 concentration every 30 seconds.

- **Subscribers**: These devices receive the requested sensor data from the brokers. Considering the preceding example, an air purifier can be a subscriber of our air quality monitoring system. It constantly receives the CO_2 concentration values, and when it crosses a threshold value, the purifier automatically turns on.

- **Broker**: This intermediary device connects various publishers and subscribers by managing and routing the data. We will be using Raspberry Pi as a broker for the entirety of this book.

Please note that both the publishers and subscribers are referred to as **clients**. A client can be a publisher, subscriber, or both as both these processes are entirely independent of each other, as we will see in the later chapters.

But another question arises now: how does the broker manage or route which information is sent where? The following section will answer this question, exploring MQTT functionality.

Functionality and components of MQTT

We have already seen the significant components of MQTT, but we will now explore how these components communicate with each other.

MQTT has no client device addresses or identifiers, making it easy to build an expansible, ad hoc network. The only thing all clients must know is the address of the broker. So, how do messages get routed between the clients? The solution for this is **topics** and **messages**.

This is how the whole system works:

1. First, the publisher sends the data collected to the broker on a particular topic, which is similar to a channel for data transmission and reception. Please note that a topic can have several subtopics too. For example, in an application where you send the temperature data from a sensor connected to your fridge, the topic will look something like this:

    ```
    Kitchen/Fridge/
    ```

    The main topic is the kitchen, and the appliance is the subtopic. The message will be `Temperature:14` on the given topic.

2. The subscribers listen to the topic. So, if the subscriber is listening to the `Kitchen` topic, it will have access to all the subtopics that are a part of this topic.

3. The primary function of the broker is to manage all the available topics and route the information according to the type of client, namely publishers and subscribers.

Now that we are aware of the details of MQTT, we will have a look at the salient features of this communication protocol.

Salient features of MQTT

This section will cover the main features of this communication protocol:

-   **Lightweight and efficient**:

    MQTT clients are tiny, and they require minimal resources to operate. So, even microcontrollers such as ESP8266 can be used as a client as long as they have an active connection to a network.

    This protocol is highly efficient thanks to the small message headers that provide maximum network bandwidth efficiency.

-   **Bidirectional communication protocol**:

    MQTT allows to-and-fro messaging capability. This means a device can be a publisher and a subscriber simultaneously. This also allows easy broadcasting of messages to several devices at once.

- **Highly scalable**:

  There is no worry about maintaining clients' addresses or IDs; it is effortless to expand the MQTT network. Moreover, the decoupling between the publishers and subscribers makes things even more accessible. The only things required on the client side are the **broker's IP address** and the **topic name**.

- **Reliability**:

  MQTT is highly reliable when it comes to message delivery. As this is an essential aspect of any communication protocol, MQTT comes with three predefined **quality of service (QoS)** levels:

  - **QoS 0**: At most once

  - **QoS 1**: At least once

  - **QoS 2**: Exactly once

- **Support for unreliable networks**:

  Many IoT devices are connected over unreliable networks, and MQTT's support for persistent sessions reduces the client's time with the broker. For example, several monitoring devices are deployed on moving vehicles or in remote areas such as forests.

- **Highly secure**:

  MQTT makes it easy to encrypt messages using TLS and authenticate clients using modern authentication protocols, such as OAuth.

This is the end of this section. We covered the basics of MQTT and the components and salient features of this popular communication protocol. This protocol will be discussed in detail in *Chapter 2, MQTT in Detail*, of this book.

A gentle introduction to Raspberry Pi

This section will introduce you to the **Raspberry Pi**, the world's most affordable credit card-sized computer.

There is a wide range of available Raspberry Pi development boards available. They are primarily available in four formats:

- **Model B**: These are full-size boards equipped with Ethernet and USB ports.

- **Model A**: These are square-shaped boards, considered *light* models of Raspberry Pi. They are different from the Model B because of the absence of an Ethernet port, fewer USB ports, and a slightly less powerful processor chip. They come at a lower price due to these cuts.

- **Zero**: This is the cheapest and smallest Raspberry Pi available. It is equipped with a significantly less powerful and low-power processor, includes no USB or Ethernet port, and is equipped with a mini-HDMI port instead of a full-size HDMI.

- **Compute**: This is Raspberry Pi 4 in a compact package for embedded applications. Additional RAM and eMMC Flash customizations are available (32 different variant configurations are listed on the official Raspberry Pi website).

The latest models of the Pi available are as follows:

- **Raspberry Pi Model 4B**

- **Raspberry Pi Model 3** (B+, B, and A)

- **Raspberry Pi Zero W**

- **Raspberry Pi 400** (a personal computer kit)

- **Raspberry Pi 4 Compute Module**

We will cover the **Raspberry Pi Model 4B** in depth as it is the latest variant available and is the model we will be using throughout the book.

Raspberry Pi Model 4B

This is the latest development board from Raspberry Pi (*Figure 1.2*). It has several new and improved features that make it an incredible upgrade over the older models. The most significant change is the support of two 4K displays, which is an astonishing feat on hardware that costs 35 dollars.

Another distinctive feature is the upgraded CPU and RAM options. The latest board is powered by a new 1.5 GHz quad-core CPU, almost three times faster than the previous-generation processor. Moreover, the boards are available in 2 GB, 4 GB, and 8 GB LPDDR4 RAM configurations.

It also has USB C support, USB 3.0 support, and Gigabit Ethernet. The Raspberry Pi 4 is a viable dual-display desktop computer with these new hardware capabilities.

Figure 1.2 – Raspberry Pi Model 4B+: the latest Raspberry Pi development board

Now, we will dig a bit deeper and cover the hardware specifications of this development board in detail followed by a brief discussion of some popular operating systems that are available for this board.

Hardware specifications

The hardware specifications of the Raspberry Pi model 4 are as follows, as mentioned on the official Raspberry Pi 4 product page:

- Broadcom BCM2711, quad-core Cortex-A72 (ARM v8) 64-bit SoC @ 1.5GHz

- 2 GB, 4 GB, or 8 GB LPDDR4-3200 SDRAM (depending on model)

- 2.4 GHz and 5.0 GHz IEEE 802.11ac wireless, Bluetooth 5.0, BLE

- Gigabit Ethernet

- Two USB 3.0 ports; two USB 2.0 ports

- Raspberry Pi standard 40-pin GPIO header (fully backward compatible with previous boards)

- Two micro-HDMI ports (up to 4kp60 supported)

- Two-lane MIPI DSI display port

- Two-lane MIPI CSI camera port

- Four-pole stereo audio and composite video port

- H.265 (4kp60 decode), H264 (1080p60 decode, 1080p30 encode)

- OpenGL ES 3.0 graphics

- microSD card slot for loading operating system and data storage

- 5V DC via USB-C connector (minimum 3A*)

- 5V DC via GPIO header (minimum 3A*)

- **Power over Ethernet** (**PoE**) enabled (requires separate PoE HAT)

- Operating temperature: 0–50 degrees °C ambient

The following figure shows the available ports and some technical specifications of the Raspberry Pi 4:

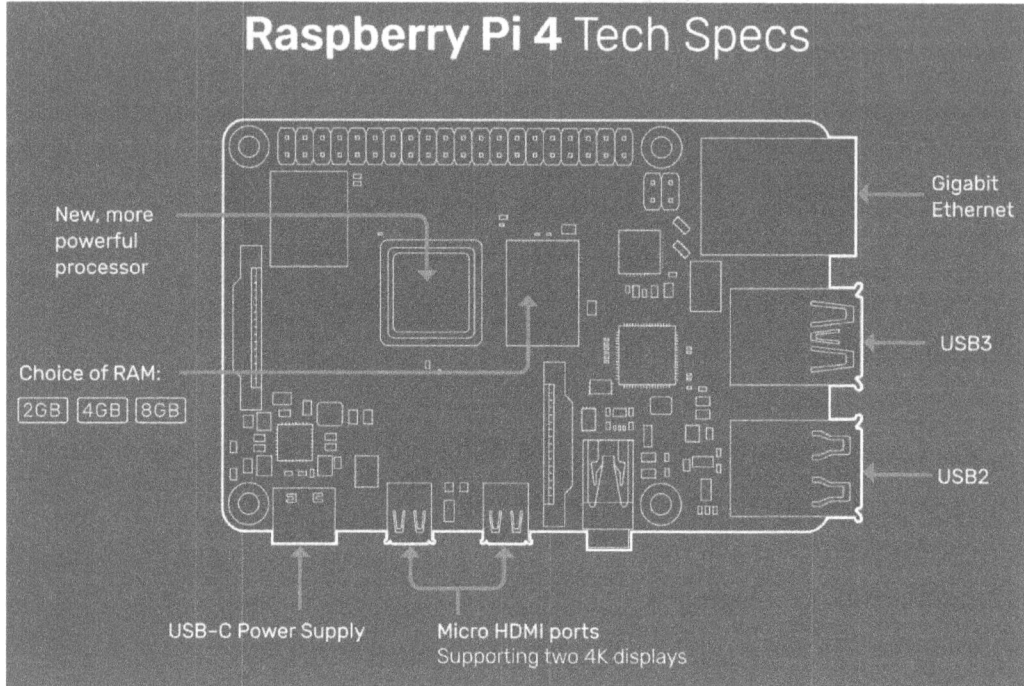

Figure 1.3 – Raspberry Pi 4 ports and hardware specifications

Now that we are done with the hardware specifications, let's move toward the available software options. As the Pi is a full-blown computer, it will run an operating system of its own. Hence, we will look at some popular operating systems available for the Raspberry Pi.

Operating systems

There are several operating systems available for the Raspberry Pi. We will look at some of the most popular operating systems listed on their official website:

- **Raspberry Pi OS** (previously known as Raspbian OS).
- **Ubuntu Core**: Ubuntu operating system developed explicitly for embedded boards, with optimizations focused on security and reliability.
- **LibreELEC**: A distribution for multimedia applications based on the Kodi entertainment center.
- **Ubuntu Desktop**: This is the desktop version of Ubuntu supported on Raspberry Pi Model 3B+ and above. One of the most popular Linux operating systems used worldwide focused on daily applications for home, school, and work.

Now that we have some knowledge about Raspberry Pi and MQTT, the next step is to learn how to setup our Raspberry Pi so that we can use it as a MQTT broker. That is exactly what the next section is about!

Setting Up Your Raspberry Pi

In this section, we will cover how to set up the **Raspberry Pi**. Although simple, it is crucial to perform each step as this will help us set up our own **local MQTT broker** on the Raspberry Pi. This will also help you understand how to get started with your new Raspberry Pi by installing an operating system onto it.

First, we will flash the popular Debian-based Raspberry Pi OS on our Raspberry Pi. After that, we will install all the necessary libraries and packages to make our device a local MQTT broker.

The topics we will be covering will be as follows:

- Setting up an SD card for your Raspberry Pi
- Flashing the OS image onto the SD card
- Setting up your Raspberry Pi for the first time
- Setting up VNC for the Raspberry Pi
- Setting up and testing the MQTT broker
- Testing the MQTT broker locally

First, we will discuss what will be required in terms of hardware to follow this setup process.

Technical requirements

To follow the instructions provided in this section, you will need the following hardware:

- Raspberry Pi (model 3B or higher, preferably Raspberry Pi 4)
- HDMI display (for first boot only)
- Keyboard and mouse (for first boot only)
- Power supply for the Pi (the official Pi power supply is recommended)
- MicroSD card (minimum 8 GB storage option is recommended)

So, let us proceed to the next step, which is installing the official Raspberry Pi OS image and setting up the SD card.

Setting up an SD card for your Raspberry Pi

In this step, the main aim is to get the microSD Card ready for the Raspberry Pi.

> **Important Note**
>
> If you have purchased a Raspberry Pi bundle with a pre-burnt SD card and some optional accessories, you can skip this step, as the SD card you have is already loaded with the required OS. But, if you would like to install a different OS, you can follow this step to do so.

Before installing the OS onto the SD card, we need to format the SD card to make sure nothing corrupts the OS. There are two methods to do so. The first is the easy way, using software to do this task, and this can be done in Windows and macOS systems. The second method is a little more complicated, and it will cover how to do the same for Linux-based systems.

The **SD Card Formatter** software (managed by the SD Association) helps you wipe the SD card totally so it can be used for the desired purpose, which, in our case, is to burn an OS image onto it.

We need to perform this step when the SD card we are using has been previously used or has some data stored on it already. This can corrupt the OS, and so all the existing data needs to be wiped. Moreover, this is perfect practice and should be done every time we install a new OS.

Let's look at the following steps:

1. To install this software, go to the relevant link depending on the OS you are using:

   - For Windows: `https://www.sdcard.org/downloads/formatter/sd-memory-card-formatter-for-windows-download/`

   - For macOS: `https://www.sdcard.org/downloads/formatter/sd-memory-card-formatter-for-mac-download/`

2. When you open the link on your browser, you will see an agreement on your screen, as shown in *Figure 1.4*. Scroll down to the end and press the **Accept** button:

Figure 1.4 – Download page for SD Card Formatter

3. Once you've clicked the **Accept** button, the software setup ZIP file will automatically start downloading on your system. Once complete, extract the ZIP file and just run the setup file. This will open an installer window; follow the steps to install the software onto your system. After the software is successfully installed, you will see the dialog box shown in the following screenshot:

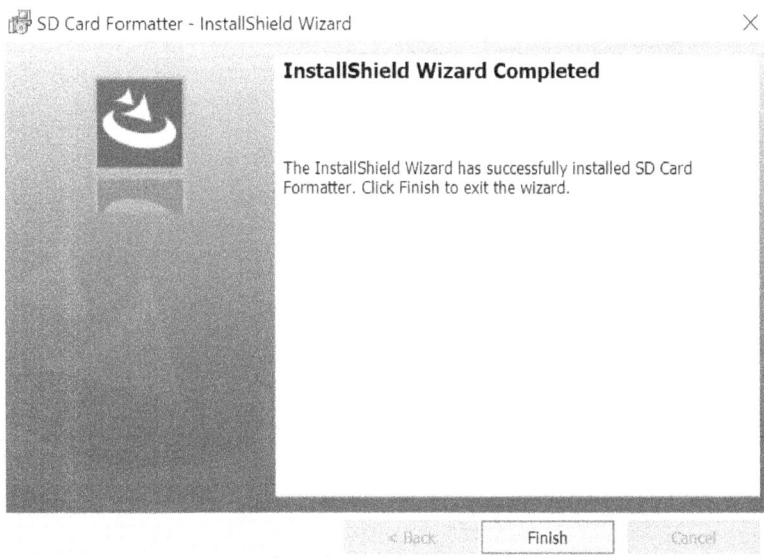

Figure 1.5 – The dialog box that appears after successful installation of the software

4. After the software has been installed, the next step is to format our SD card. For this, an SD card reader is required. There are two options available on the market. One is a USB SD card reader stick, and the other is an SD card adapter, as shown in the following figure:

Figure 1.6 – Common SD card adapters

5. Once getting an adapter, the next step is to insert the SD card into the adapter and then insert it into the PC. Please note that not all PCs and laptops have SD card readers, so getting a USB adapter is better to avoid any problems.

6. Once the stick has been inserted and detected by your machine, launch the SD Card Formatter software. A dialog box will open, which looks something like the following:

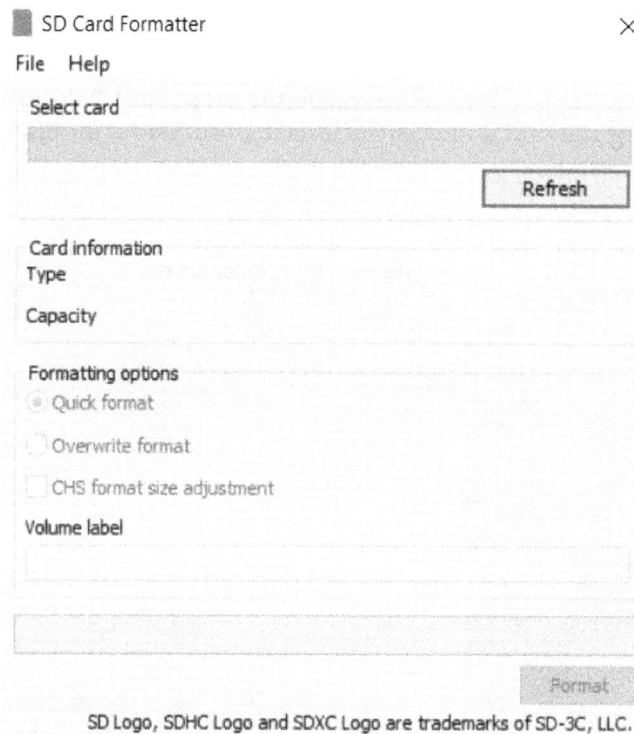

Figure 1.7 – SD Card Formatter application page

7. Next, you have to select your card, which will be visible in the **Select card** dropdown menu. There will be two partitions available (in your file explorer) for formatting if you have an OS image already burned on it. In that case, select the one that has the name **boot**.

8. Once the card is selected, keep all the other settings as default and press the **Format** button. This will start the process of wiping the SD card. Once the process is complete, you will see a dialog box saying **Formatting was successfully completed**, as seen in the following screenshot:

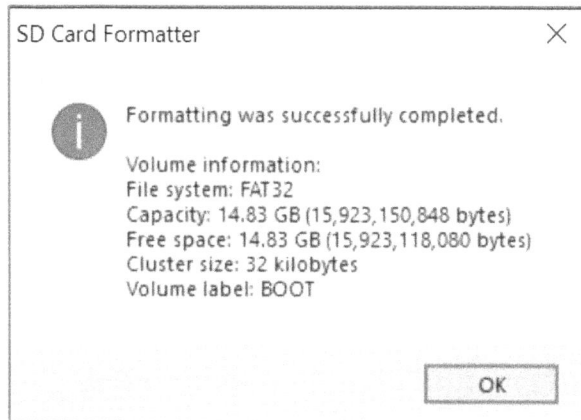

Figure 1.8 – Formatting completion dialog box

We have successfully formatted our SD card! Now, we are all set to burn our Raspberry Pi OS image onto our card.

> **Important Note**
>
> Please keep in mind that if you are using an SD card adapter, make sure it is in *unlock* mode or you could face formatting issues. Unlocking the adapter means giving the computer access to the SD card. This is done by simply flipping a small switch on the side of the adapter.

SD card formatting in Linux systems

We will use **GParted** to format our SD card on a Linux system. It is an open source disk management software. Just follow these steps:

1. First, we will need to install this software, as this does not come preinstalled. We will use the Ubuntu OS for this tutorial, the most common and widely used Linux OS. Use this command to install the software via the Linux Terminal:

```
sudo apt install gparted
```

Once the app has been installed, it will be available in the **Applications** menu. Just find and launch the application, as shown in the following figure:

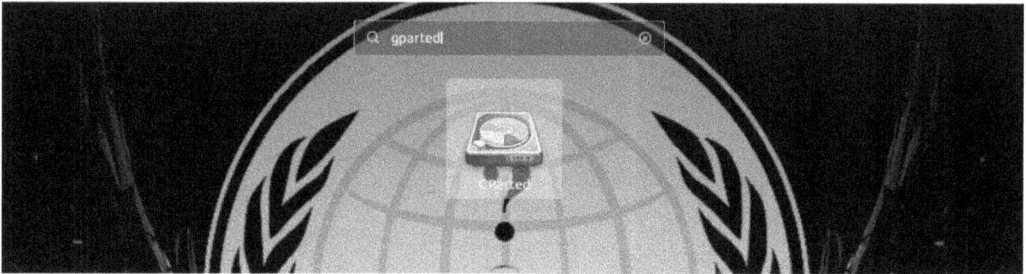

Figure 1.9 – Launching GParted from the Applications menu

2. This application requires root privileges to run, so enter your password when prompted. After that, the application window will pop open, and you will be able to see all the disks presently connected to your system:

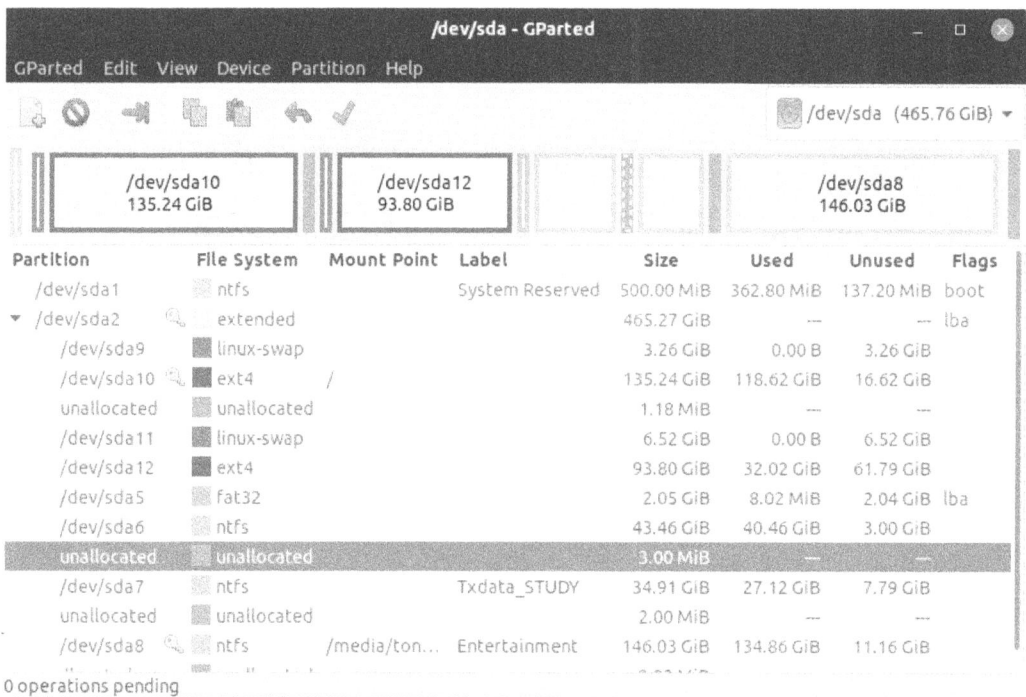

Figure 1.10 – GParted application home screen

The user interface is relatively easy to navigate and much more intuitive than the one we get with the preinstalled *disk utility software*. There are additional features that this software provides, such as creating bootable USB drives and downloading an ISO file, for example.

3. Now, select the SD card drive from the top-right corner dropdown, as shown here:

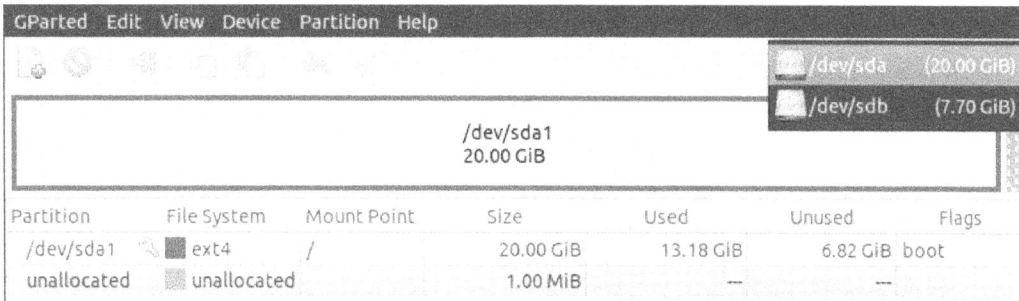

Figure 1.11 – Selecting the USB drive to format

4. We will now format this drive, but to do so, we first need to unmount it. Just right-click on the visible partition, and you will see an option to unmount it , as shown in the following screenshot:

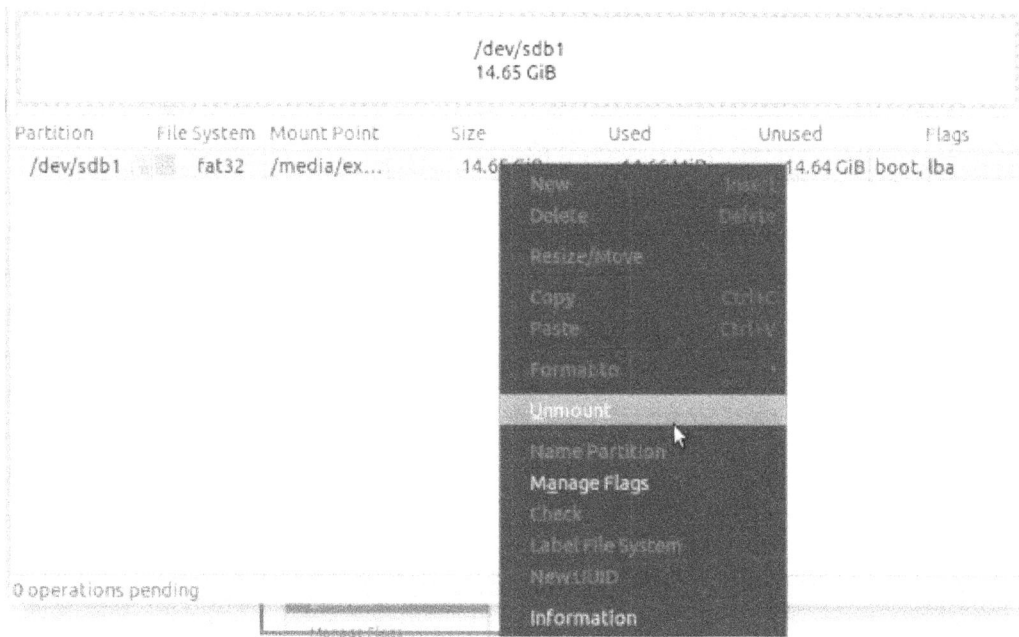

Figure 1.12 – Unmounting the USB adapter before formatting

5. You can start the formatting process once the drive has been unmounted. To do that, right-click on the USB drive and select the **Format to** option. Select the file system of your SD card (**fat32**, in most commonly available SD cards):

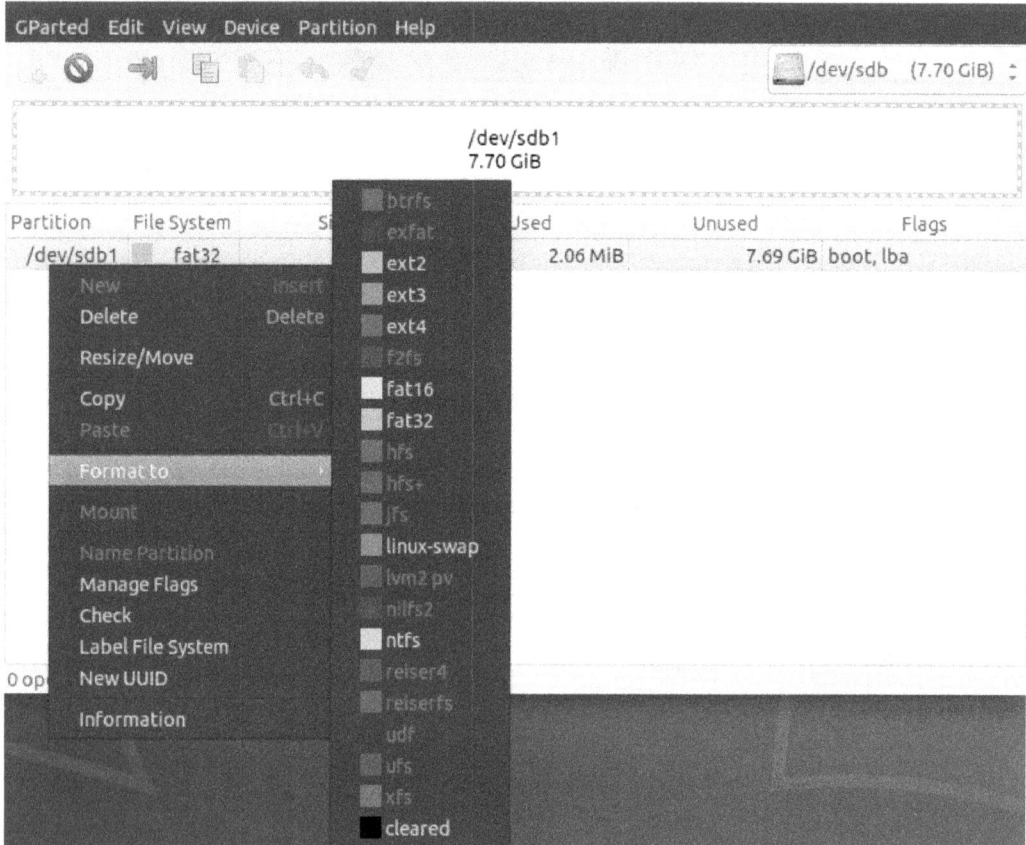

Figure 1.13 – Selecting the desired file format (fat32 in most cases)

This won't start the formatting process. It will just add a new operation to the list of pending operations, as seen here:

Figure 1.14 – List of pending operations

6. Now, click the **Apply All Operations** button, which is the green tick icon at the top, as shown in *Figure 1.15*:

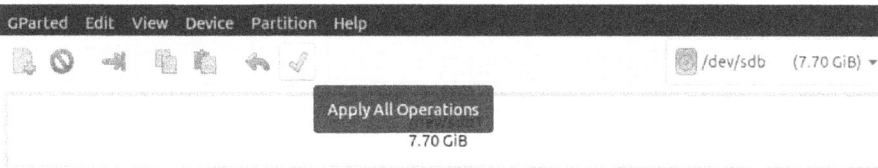

Figure 1.15 – Clicking on Apply All Operations

7. A window will pop up, warning you that the operation will lead to complete loss of data on the USB drive. Just click on **Apply** to start the formatting process:

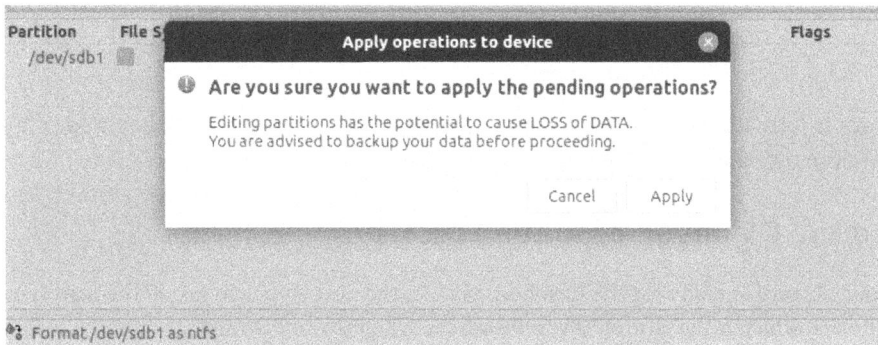

Figure 1.16 – Clicking Apply to continue formatting

The formatting process will start. You can track the progress using the window that pops up:

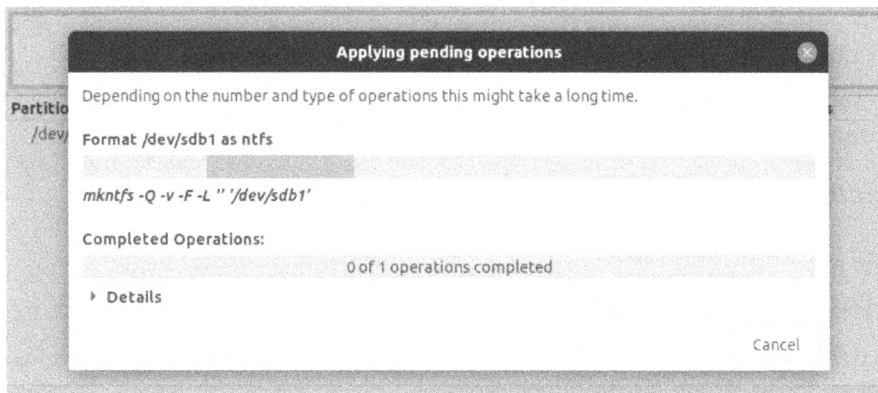

Figure 1.17 – Formatting progress dialog box – GParted

You will see the window shown in the following screenshot once the formatting process has been completed:

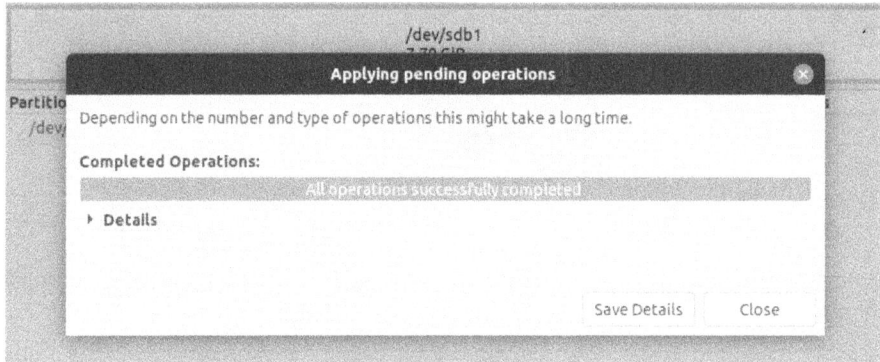

Figure 1.18 – Formatting process completed dialog box

Your SD card is formatted! Close the application, and your USB adapter will be listed in the file manager. We can now move to the next step.

Flashing the OS image onto the SD card

Now that our SD card is ready for the Raspberry Pi OS, the next step is to install the **Raspberry Pi Imager** software, which is the easiest way to install the OS onto our SD card.

> **Important Note**
>
> For more advanced users who are looking to install a particular OS, follow this link: `https://www.raspberrypi.org/documentation/installation/installing-images/README.md`.

In the next section, we're going to follow the step-by-step process to install the Raspberry Pi OS on your SD card.

Downloading and installing the Raspberry Pi Imager software

The first step is to install the software:

1. To do that, visit the following link: `https://www.raspberrypi.org/software/`.

2. Once the page is loaded, you will see a section for Raspberry Pi Imager. Just download the latest version of the software for your OS:

Install Raspberry Pi OS using Raspberry Pi Imager

Raspberry Pi Imager is the quick and easy way to install Raspberry Pi OS and other operating systems to a microSD card, ready to use with your Raspberry Pi. Watch our 40-second video to learn how to install an operating system using Raspberry Pi Imager.

Download and install Raspberry Pi Imager to a computer with an SD card reader. Put the SD card you'll use with your Raspberry Pi into the reader and run Raspberry Pi Imager.

Download for Windows

Download for macOS

Download for Ubuntu for x86

Figure 1.19 – Downloading Raspberry Pi Imager

3. Once you press any one of the download links, the latest version of the installer will download onto your system. (At the time of writing this book, the latest version is *v1.5*.)

4. To run the installer, follow the process by pressing the **Next** button until the software is installed. Once the installation is complete, open the Raspberry Pi Imager software. A dialog box will pop up asking for permission as it requires root/administrator access. Just allow it, and the application will open:

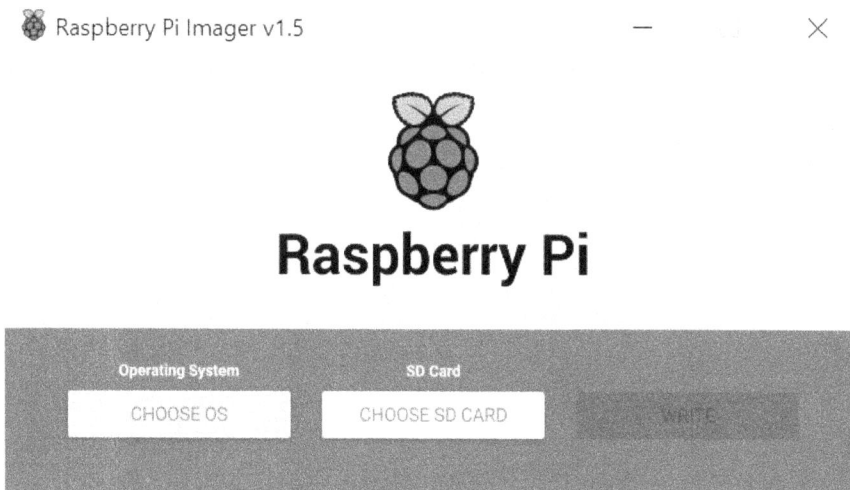

Raspberry Pi Imager v1.5

Raspberry Pi

| Operating System | SD Card | |
| CHOOSE OS | CHOOSE SD CARD | WRITE |

Figure 1.20 – Raspberry Pi Imager application

5. We have to choose the OS we want to install, which is the Raspberry Pi OS. To do that, press the **CHOOSE OS** button, and you will see all the available options as follows:

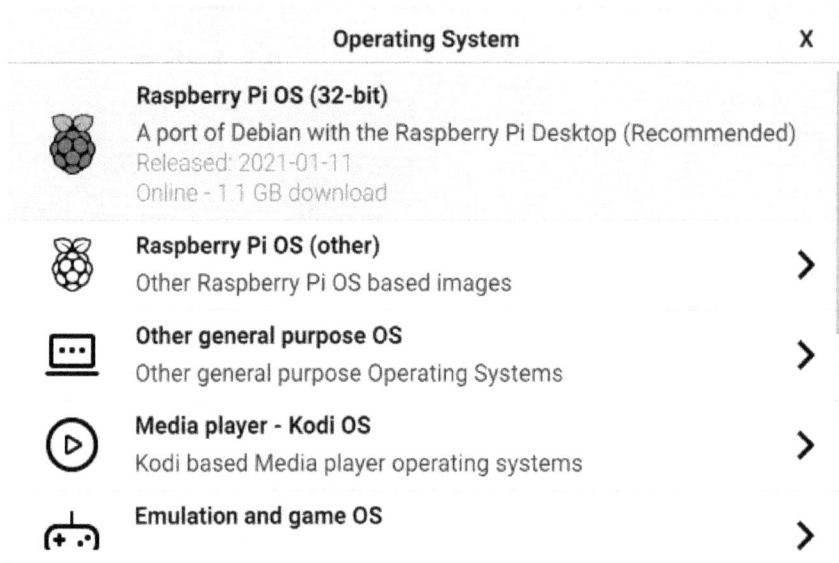

	Operating System	X
	Raspberry Pi OS (32-bit)	
	A port of Debian with the Raspberry Pi Desktop (Recommended)	
	Released: 2021-01-11	
	Online - 1.1 GB download	
	Raspberry Pi OS (other)	>
	Other Raspberry Pi OS based images	
	Other general purpose OS	>
	Other general purpose Operating Systems	
	Media player - Kodi OS	>
	Kodi based Media player operating systems	
	Emulation and game OS	>

Figure 1.21 – All the available OS options

6. We will select the first option, **Raspberry Pi OS (32-bit)**. As seen in *Figure 1.21*, the size of the latest OS version is **1.1 GB**. That means that before burning the OS, the system will download the 1.1 GB image file.

7. Next, choose the SD card directory you want to burn the OS onto. At this point, insert your SD adapter if you haven't already, and you will be able to see it listed when you press the **CHOOSE STORAGE** button:

	SD Card	X
	USB Mass Storage Device USB Device - 15.9 GB	
	Mounted as F:\	

Figure 1.22 – Choosing the SD card you want to burn to OS to

8. After both the OS and SD card have been selected, press the **WRITE** button, which will now be white-colored:

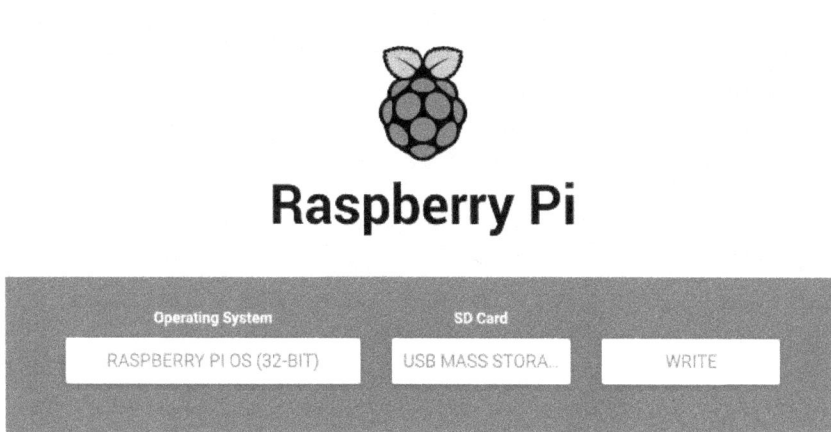

Figure 1.23 – Pressing the Write button

This will start the writing process. The application will first download the image file you chose and then write it onto the SD card. Please note that you will not see the download progress, just a **Writing** progress bar. So, it is an excellent time to grab a cup of coffee or go for a short walk, as this may take some time:

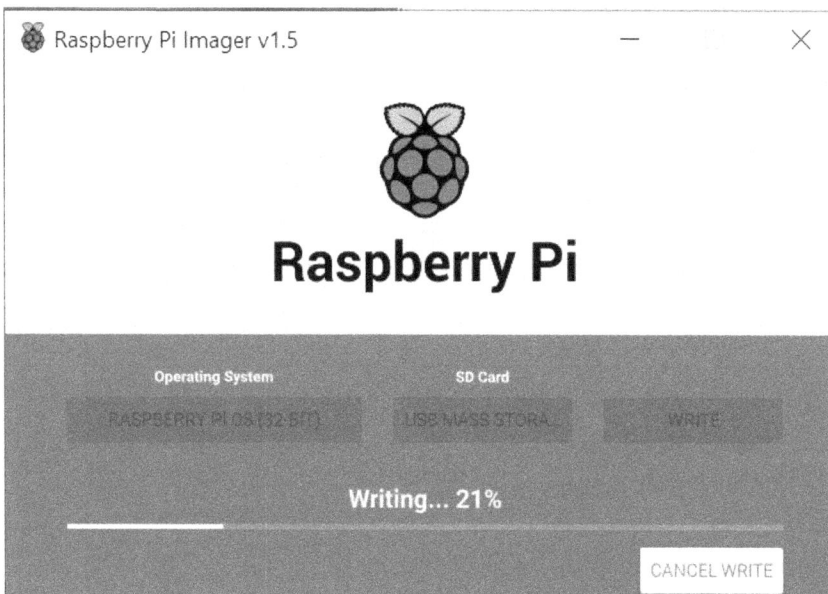

Figure 1.24 – You can see the writing progress in the application

9. Once the writing process is over, you will see a dialog box saying **Raspberry Pi OS has been successfully written…**. You can now press the **Continue** button and remove the adapter.

This completes the SD card preparation for our Raspberry Pi. In the next step, we will boot into our new OS for the first time and update and upgrade some software to the latest version, enabling VNC to wirelessly SSH into our Pi (don't worry, we will discuss this in detail in a later section).

Setting up Raspberry Pi for the first time

After the OS has been written to the SD card, we will insert this card into our Raspberry Pi, as seen in the following figure:

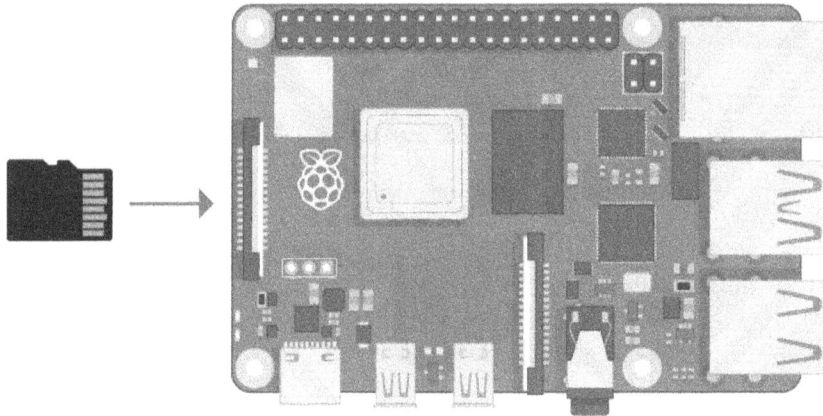

Figure 1.25 – Inserting the SD card into the Pi (image from the official Raspberry Pi website)

We will also connect a display using either a display port (if you have a Raspberry Pi 4) or a simple HDMI display, and a USB keyboard and mouse. Finally, after all the peripherals are connected, we will connect our power supply (any USB C or micro-USB charger, depending on the model you are using) to the Pi. Please see the following figure for how to make the connections:

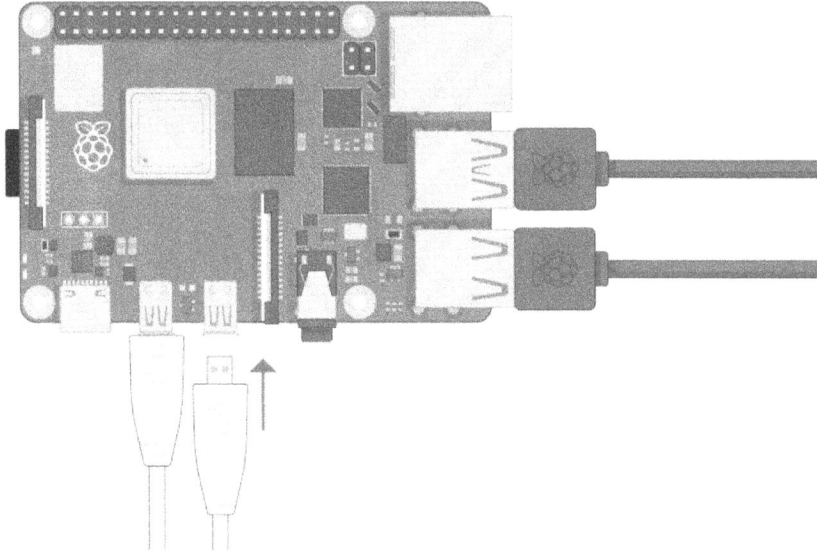

Figure 1.26 – Powering the Raspberry Pi after connecting the peripherals

Once all the setup is complete, connect the power supply to the Pi, and you should see the Pi booting up on the connected HDMI screen. It will take 20-30 seconds for the first boot. Once it is done, a welcome screen will appear saying **Welcome to Raspberry Pi Desktop**, as seen in *Figure 1.27*:

Figure 1.27 – Raspberry Pi welcome screen

Just complete the first-time setup by clicking on the **Next** button. It will first ask you to set the location settings and choose the language and keyboard accordingly:

Figure 1.28 – Setting up the location, language, and timezone

After the required information has been entered, press the **Next** button. It will take a few seconds for the system to set up the location.

Next, the OS will prompt you to change the default password of your system, which is **raspberry**. Select a strong password, and after entering all the required information, press the **Next** button:

Figure 1.29 – Setting up a new password

After this, the system will help you choose the best resolution according to your display type. You can skip this step, as we will only use this monitor or display once. In the next section, we will learn how to set up VNC and SSH on our Pi to access it wirelessly when we are connected to the same network.

Next, we need to connect to a Wi-Fi network, as shown in *Figure 1.30*. Select your network from the list of available networks, then authenticate by entering your password to connect to your network. If you have connected via Ethernet or would like to do it later, you can skip this step by pressing the **Skip** button.

Please note that you will require a shared Wi-Fi network to use VNC, and you will need a local network at the very least. If you skip the step, for now, you can always connect to a network through the Wi-Fi symbol on the top-right side of the desktop in the following figure:

Figure 1.30 – Connecting to a Wi-Fi network

Finally, the last step of the setup is to update the software to the latest version. As we have used the newest version of Raspberry Pi OS, no significant updates will be pending. It is still preferable to use this opportunity to update all the preinstalled software to the latest version (*Figure 1.31*).

Just click **Next** as given in the instructions, and the process will automatically start:

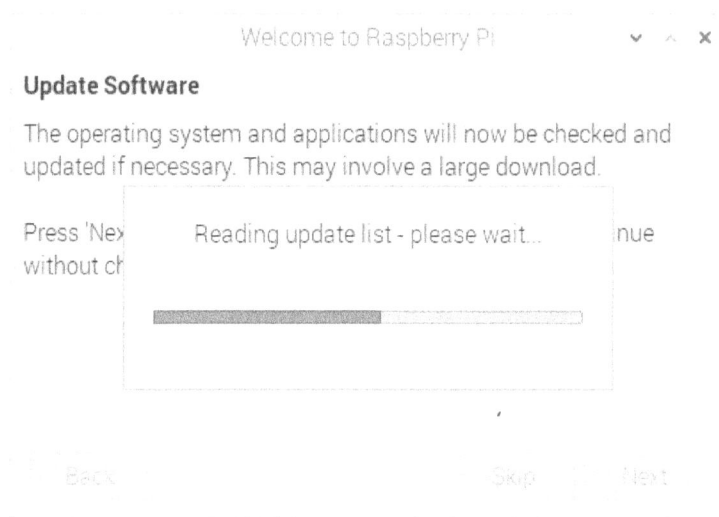

Welcome to Raspberry Pi   ∨ ∧ ✕

Update Software

The operating system and applications will now be checked and
updated if necessary. This may involve a large download.

Press 'Nex   Reading update list - please wait...   nue
without ch

Back   Skip   Next

Figure 1.31 – Updating to the latest software

> **Important Note**
>
> If you fail to update your system through this setup, it is possible to do this later. For this,
> open the terminal (the black icon on the top-left side of the home screen) and type the
> following commands:
>
> `sudo apt update` – This will fetch the list of all available updates.
>
> `sudo apt dist-upgrade` – This will download and install the updates.

It will take some time to fetch all the updates and install them, so this would be an excellent time to
go for a short walk.

After the updates have been successfully downloaded and installed, the system will prompt you to
restart the system for all the changes to take effect. You can do so by simply pressing the **Restart**
button, as shown in the following figure:

Figure 1.32 – Restarting the system after the first setup

This completes the first-time setup of our Raspberry Pi! In the next section, we will learn how to set up VNC on our Raspberry Pi to access it wirelessly through our PC. What's more, we can also use our PC or laptop keyboard and mouse with the Pi.

Setting up VNC for Raspberry Pi

In this step, we will learn how to set up **SSH** (**Secure Shell**) and **VNC** (**Virtual Network Computing**) on the Raspberry Pi. Note that the completion of the previous steps of the OS installation and setup on the Raspberry Pi is required to set these up.

Before getting into the practical part, let's learn a bit more about VNC and SSH:

- **SSH**: SSH is a security protocol that gives you remote access to your computer. It can be used for both remote login and file transfer. It provides several alternating options for strong authentication. The following figure shows how SSH technology actually works:

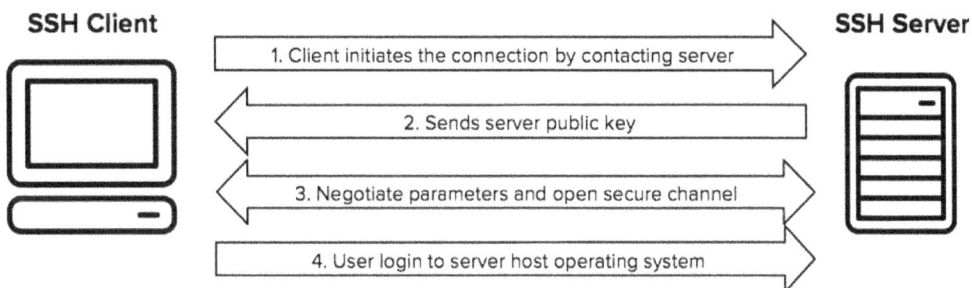

Figure 1.33 – How does SSH work?

- **VNC**: This is a cross-platform desktop sharing system that allows you to remotely access any computer system with a server through supported client software. It uses the **Remote Frame Buffer** (**RFB**) protocol to achieve this. It even enables the use of the keyboard and mouse of the client system as it relays all this information to and fro over a network.

We will be using the **RealVNC Server** software, which is preinstalled on our Pi. Raspberry Pi will act as a VNC server, and a client software would be needed to access it through the main computer.

Let's get started on the steps to set up VNC:

1. As mentioned earlier, the software comes preinstalled on the Raspberry Pi OS (the Desktop OS version), but it is disabled by default. These options must be enabled.

2. First, open the **Start** menu. (It is the Raspberry Pi logo on the top left of the screen.) From there, make these selections:

    Preferences | Raspberry Pi Configuration

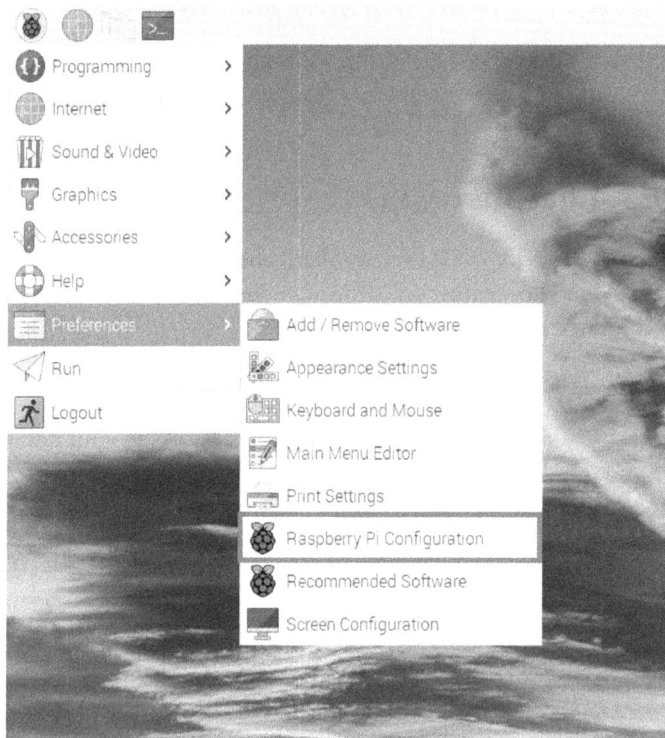

Figure 1.34 – Selecting Raspberry Pi Configuration

3. Once the **Raspberry Pi Configuration** dialog box opens, there will be five sections available, from which we need to select the **Interfaces** section:

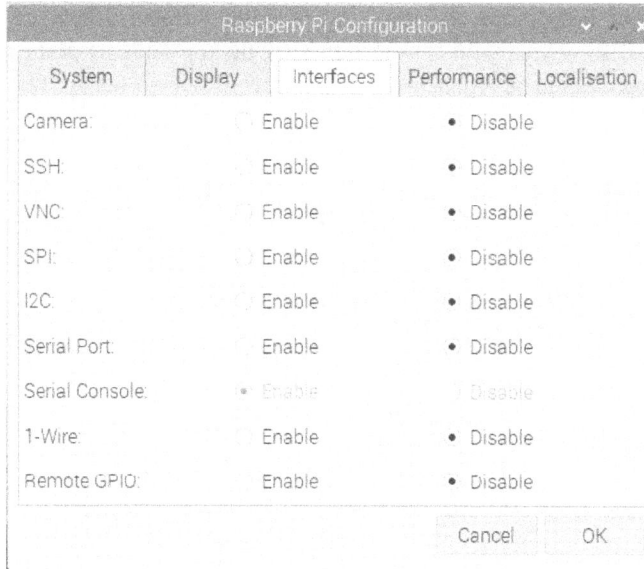

Figure 1.35 – Raspberry Pi Configuration Interfaces

4.   Enable the VNC and SSH options, then press **OK**.

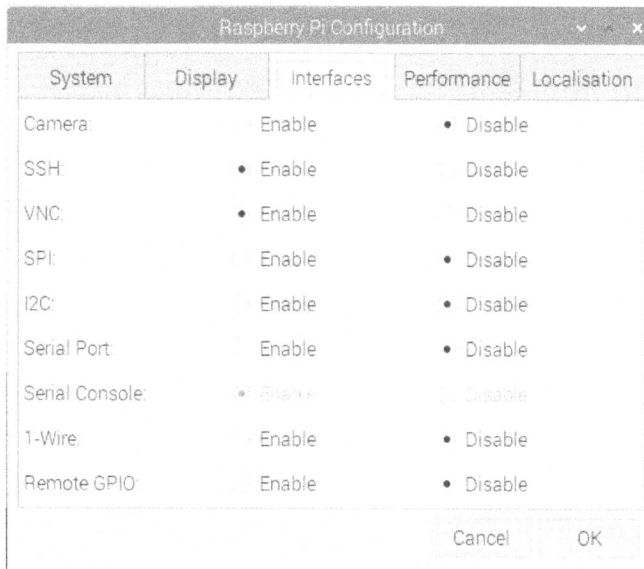

Figure 1.36 – Enabling SSH and VNC

This will enable both the protocols on your Pi. Now, the Raspberry Pi is remotely accessible. Congratulations! To verify that it is working, check the top-left part of the screen. A white-colored VNC logo should be visible now:

Figure 1.37 – Checking for the VNC logo when enabled

5. Click on the VNC icon to get the IP address of our Pi. Make a note of this, as it will be required to access your Pi remotely from your main computer:

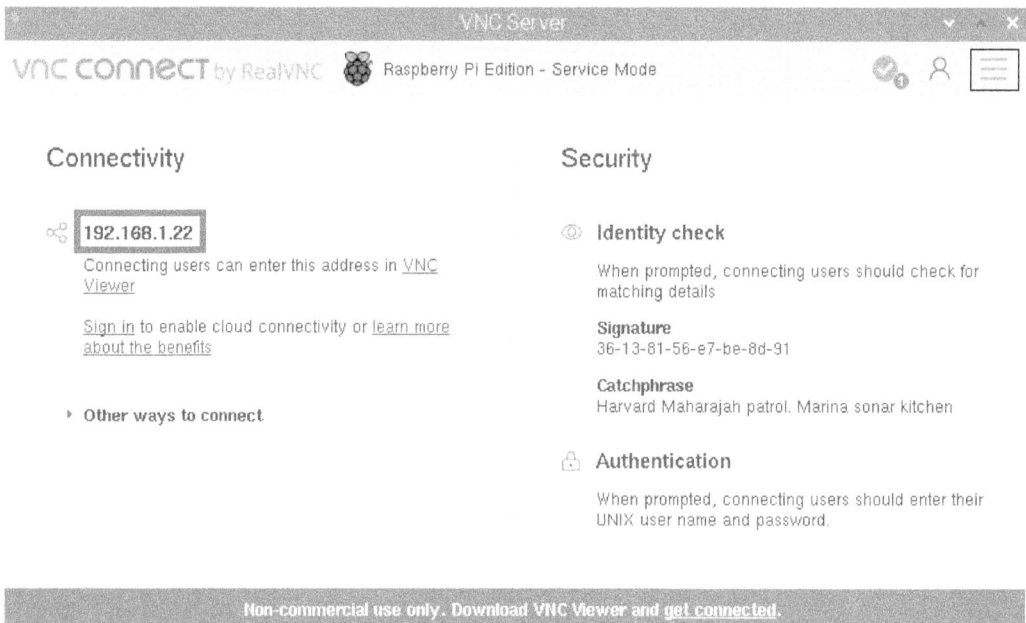

Figure 1.38 – Make a note of the IP address

6. Move on to the main computer you will be using to access the Pi. The first requirement is to download the **VNC Viewer** software. Visit the following link to download it: `https://www.realvnc.com/en/connect/download/viewer/`.

This is what the **Download** page looks like:

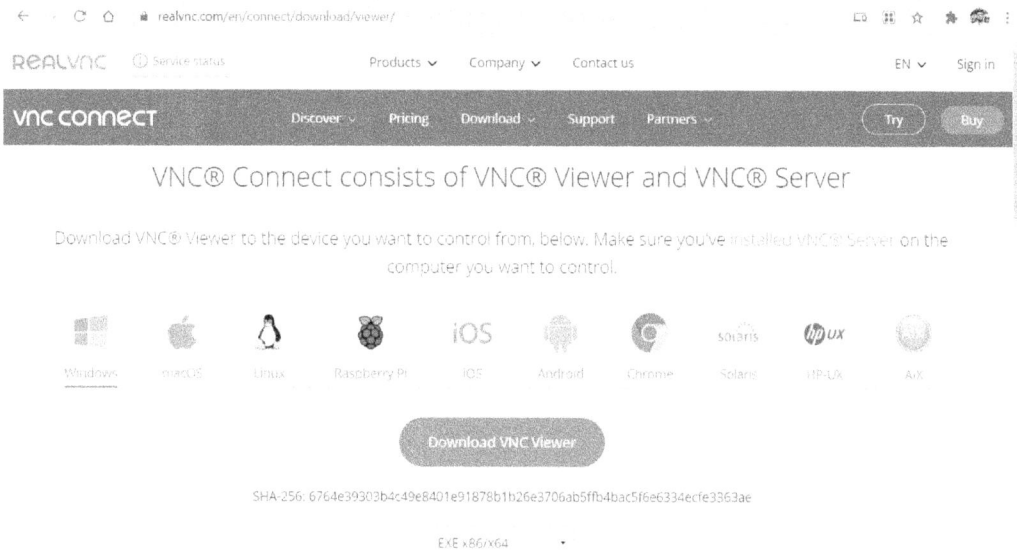

Figure 1.39 – The VNC Viewer download page

7. After downloading the installer for your specific OS, install it onto your system and open the application. The application home page will look something like this:

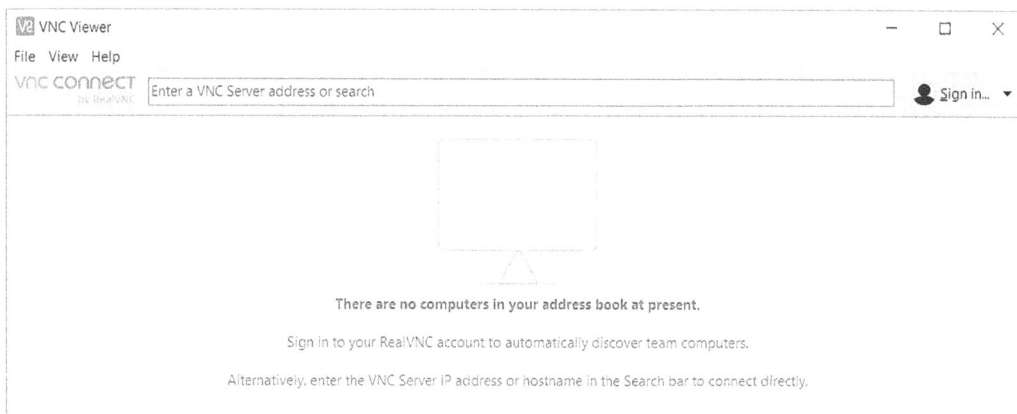

Figure 1.40 – The VNC Viewer application

8. Make sure that your Pi and computer are connected to the same network. Once that is done, enter the IP address of your Pi in the search bar of the application, and then click **Enter**. It will show a warning message; just press **Continue**, and you will then see an **Authentication** window:

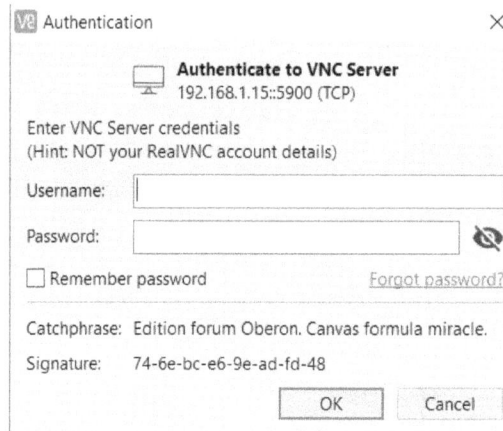

Figure 1.41 – The Authentication window

9. Enter the username and password and press **OK**. If the credentials are correct, your attempt to establish a connection will be successful, and now you can remotely log in to your Pi:

Figure 1.42 – Remote access to Raspberry Pi

You now have access to your Raspberry Pi wirelessly with the ability to use your computer mouse and keyboard with your Pi's OS.

The next step is crucial, as we will install a software library that will allow us to use our Raspberry Pi as an MQTT broker. We will install the following two libraries developed by **Mosquitto**:

- *Mosquitto – MQTT Broker Library.*

- *Mosquitto Clients* (optional) – This library allows you to run client code on the Pi.

Setting up and testing the MQTT broker

After this setup step and the testing step that follows, you will be able to host a local MQTT broker on your Raspberry Pi device. Moreover, you will also test whether your broker is successfully running by simultaneously running a client code on the Pi itself. Cool, right?

So, let's get started with the steps to get this done.

Installing the MQTT broker and Clients packages

Mosquitto is a popular MQTT broker well-supported on Debian-based Linux platforms such as Raspbian. To install this package, just open a new terminal, then type in the following command. It's easy to install using the `apt` package installer:

```
sudo apt install mosquitto mosquitto-clients
```

This command requires root privileges for which we have used `sudo` in our command. After that, the installation process will start, and once it is complete, you should see an output as shown in the following screenshot:

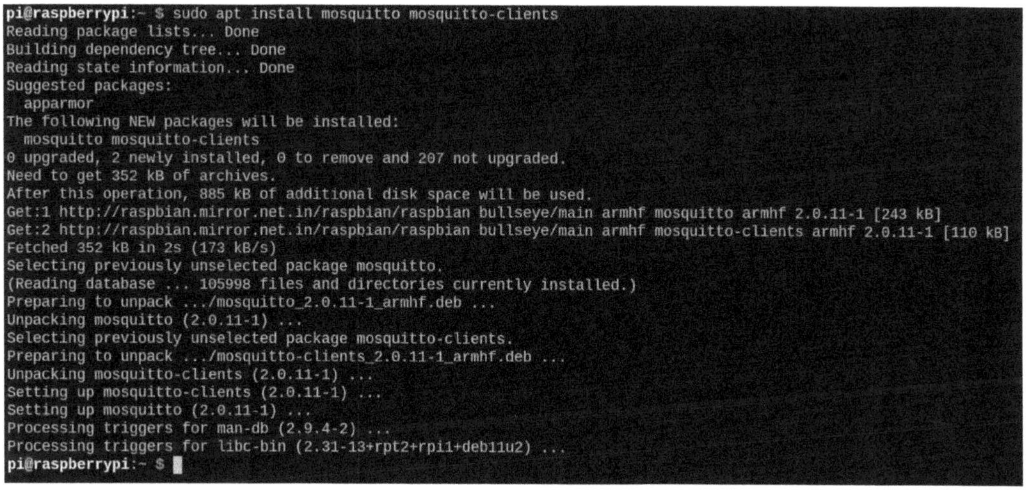

```
pi@raspberrypi:~ $ sudo apt install mosquitto mosquitto-clients
Reading package lists... Done
Building dependency tree... Done
Reading state information... Done
Suggested packages:
  apparmor
The following NEW packages will be installed:
  mosquitto mosquitto-clients
0 upgraded, 2 newly installed, 0 to remove and 207 not upgraded.
Need to get 352 kB of archives.
After this operation, 885 kB of additional disk space will be used.
Get:1 http://raspbian.mirror.net.in/raspbian/raspbian bullseye/main armhf mosquitto armhf 2.0.11-1 [243 kB]
Get:2 http://raspbian.mirror.net.in/raspbian/raspbian bullseye/main armhf mosquitto-clients armhf 2.0.11-1 [110 kB]
Fetched 352 kB in 2s (173 kB/s)
Selecting previously unselected package mosquitto.
(Reading database ... 105998 files and directories currently installed.)
Preparing to unpack .../mosquitto_2.0.11-1_armhf.deb ...
Unpacking mosquitto (2.0.11-1) ...
Selecting previously unselected package mosquitto-clients.
Preparing to unpack .../mosquitto-clients_2.0.11-1_armhf.deb ...
Unpacking mosquitto-clients (2.0.11-1) ...
Setting up mosquitto-clients (2.0.11-1) ...
Setting up mosquitto (2.0.11-1) ...
Processing triggers for man-db (2.9.4-2) ...
Processing triggers for libc-bin (2.31-13+rpt2+rpi1+deb11u2) ...
pi@raspberrypi:~ $
```

Figure 1.43 – Terminal view while installing the packages

The `mosquitto-clients` package is optional for running the Pi as an MQTT broker, but it will help us test whether the broker is running locally.

This package allows you to use your Raspberry Pi as an MQTT client as well. So, if you want to create a local dashboard to control all the MQTT clients from your Pi, you will be able to do so.

Enabling the Mosquitto broker

The broker is still not active. To enable it, type in the following command in your terminal window:

```
sudo systemctl enable mosquitto
```

`systemd` is a Linux package manager that will help you monitor and control the different applications installed. Once the command has been executed, the broker should be running on your Pi. To confirm that, just run the following command:

```
sudo systemctl status mosquitto
```

This should produce an output on the terminal window similar to what is shown in the following screenshot:

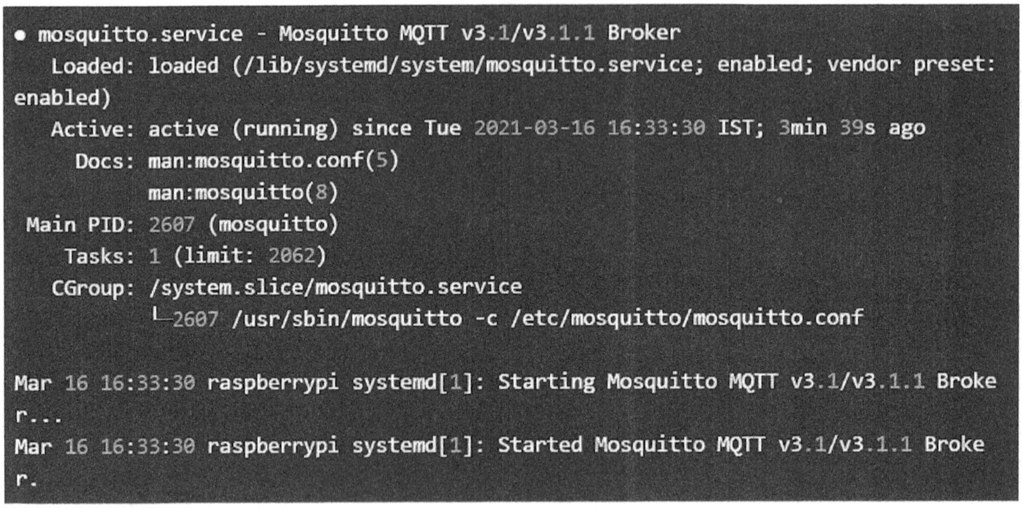

```
● mosquitto.service - Mosquitto MQTT v3.1/v3.1.1 Broker
     Loaded: loaded (/lib/systemd/system/mosquitto.service; enabled; vendor preset:
enabled)
     Active: active (running) since Tue 2021-03-16 16:33:30 IST; 3min 39s ago
       Docs: man:mosquitto.conf(5)
             man:mosquitto(8)
   Main PID: 2607 (mosquitto)
      Tasks: 1 (limit: 2062)
     CGroup: /system.slice/mosquitto.service
             └─2607 /usr/sbin/mosquitto -c /etc/mosquitto/mosquitto.conf

Mar 16 16:33:30 raspberrypi systemd[1]: Starting Mosquitto MQTT v3.1/v3.1.1 Broke
r...
Mar 16 16:33:30 raspberrypi systemd[1]: Started Mosquitto MQTT v3.1/v3.1.1 Broke
r.
```

Figure 1.44 – The output of the status command for the broker

The most important thing is that the `Active` option should show the `active (running)` status, which will verify that our broker is up and running!

> **Important Note**
>
> If the status command shows an output that says that your process is dead and your MQTT broker stopped, restart the MQTT service by typing the following command:
>
> ```
> sudo service mosquitto restart
> ```
>
> Now, recheck the status, and it should show the status of your MQTT broker as running!

This marks the conclusion of this section. We successfully set up our Raspberry Pi as a local MQTT broker. Additionally, we installed a package that will let us use the Pi as an MQTT client as well. In the next section, we will test our MQTT broker's functionality through a short demonstration.

Testing the MQTT broker locally

Now that the MQTT broker is running on the Raspberry Pi, we will test the connection using a straightforward project. First, open two terminal windows on your Raspberry Pi. Now, we will do the following:

1. In one terminal window, we will subscribe to a particular MQTT topic, for instance:

    ```
    test/message
    ```

    To do this, you will require the `mosquitto-clients` package. Now, type the following command in the terminal:

    ```
    mosquitto_sub -v -t test/message
    ```

    This will subscribe to the topic entered after `-t`.

2. Next, in the alternate terminal window, we will publish a test message on the same topic to check whether it is sent and received by the other terminal window.

    Just imagine the two terminals as different MQTT clients where one client is a subscriber to the topics which the other client publishes. To publish a test message of `'Hello World!'`, type the following command:

    ```
    mosquitto_pub -t test/message -m 'Hello World!'
    ```

3. After running the previous command on the second terminal, you should see the message `'Hello World!'` on the terminal along with the topic name, as illustrated in the following screenshots:

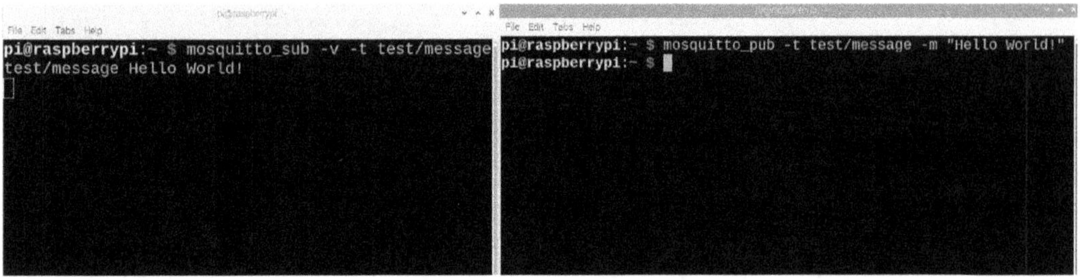

Figure 1.45 – Raspberry Pi local MQTT test

Congratulations! You have successfully set up your Raspberry Pi. This marks the end of the first chapter of this book. Now, let us summarize what we covered in this chapter.

Summary

This chapter introduced us to Raspberry Pi and MQTT. We started with a brief introduction about MQTT.

Next, we covered the Raspberry Pi, the main hardware we will be using throughout this book. We started by covering the hardware specifications and some popular operating systems available for the Pi. Then we set up our Raspberry Pi to work as a local MQTT broker. To do that, we set up the SD card we will be using for our Pi by formatting it and then flashing the latest version of Raspberry Pi onto it. After that, we did the essential steps to set up our Raspberry Pi for the very first time. Next, we installed the packages that let us use our Pi as an MQTT broker as well as an MQTT client. Finally, we tested the functionality of our Pi as a local MQTT broker.

In the next chapter, we will dive deeper into the different clients we can use with our broker. We will set up an MQTT client on our main computer to understand how MQTT actually works.

2
MQTT in Detail

This chapter touches on one of the main topics of this book: **MQTT**. As you saw in the previous chapter, when we used our Raspberry Pi as an MQTT client, it was straightforward and included just one command that needed to be executed. But the question is, how does the client connect to the broker, and how does it send the message to the intended client?

This chapter will address that exact point. It will significantly help you learn what MQTT is and how it works under the hood. You will gain a clear understanding of this communication protocol, along with the ability to use your laptops and computers as local MQTT clients, which is a bonus.

We're going to cover the following main topics in this chapter:

- Introducing MQTT clients
- Understanding the MQTT protocol packet structure
- Practical demonstration of MQTT in action

Let's go ahead and begin.

Introducing MQTT clients

MQTT communication flow consists of a client (which can be a publisher or subscriber and in certain instances, both) and the broker, which manages the flow of all information across different clients. The following diagram provides an overview of how the *MQTT message flow* works:

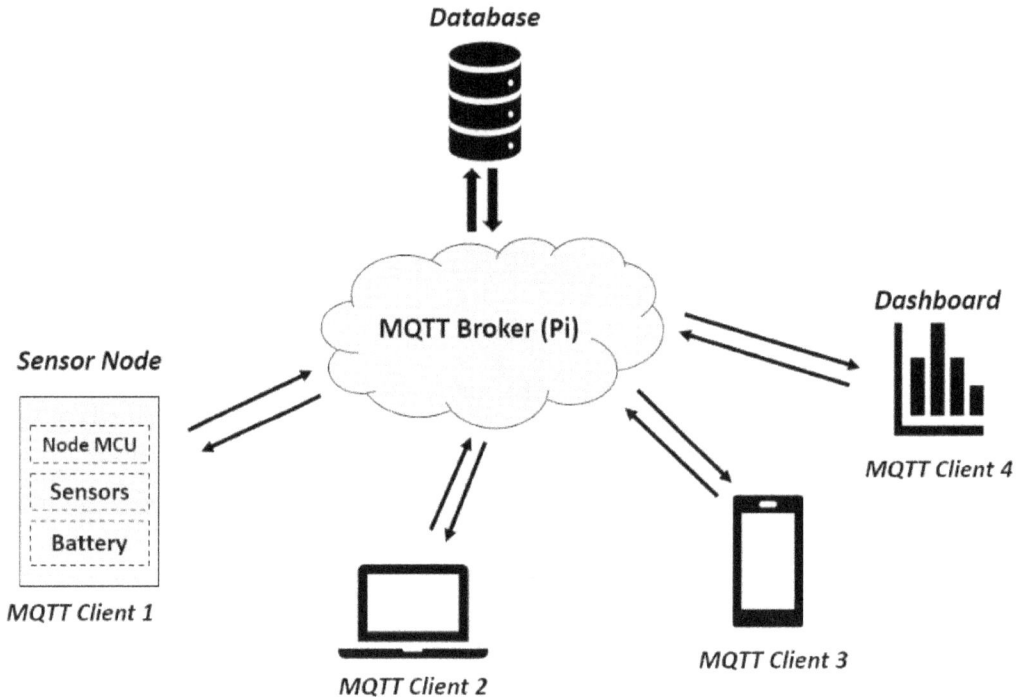

Figure 2.1 – MQTT overview

As discussed earlier in this book, MQTT stands for **Message Queuing Telemetry Transport**. Simply put, it is a communication protocol designed for constrained devices with network limitations. It is designed as a lightweight publish/subscribe messaging protocol. But what does this mean? For this, we need to be familiar with the concepts of **messages**, **topics**, **clients**, and **brokers**. Let's cover each and how they work.

MQTT messages

A message is a term given to the data that's shared between different MQTT clients. It can be some text, sensor readings, and so on.

MQTT topics

Topics are one of the essential components of this protocol. They provide you with a unique address for where your message should go. An MQTT topic is a series of strings separated by forward slashes. Each string before a forward slash indicates a new topic level. This gives you a lot of options for unique topics. Here is an example of a topic:

- `Bedroom/Lighting/Lamp`

In this example, the topic is interpreted as follows: under the main topic, `Bedroom`, there is a sub-topic called `Lighting`, and under that, there is a subtopic called `Lamp`. If we send any message to this topic via a client, any clients connected to this particular topic via our *broker* will receive this message.

There may be cases when you would want to subscribe to multiple topics from a single client. For instance, if you want to connect to 30 such topics, it would be very tedious to write each topic name. There is where **Wildcards** come into play. They let you subscribe to multiple topics with a single statement. There are two types of wildcards in MQTT: **single-level** and **multi-level**.

Single-level wildcards

In the case of single-level wildcards, you can use them to substitute a single sub-topic hierarchy. This can be done by simply using the + symbol instead of the subtopic's name:

Bedroom/+/Lamp

In this case, you can create any topic with the main topic as `Bedroom` and the third subtopic as `Lamp`. The value of the second topic can be anything:

- `Bedroom/Lighting/Lamp`
- `Bedroom/State/Lamp`

However, you can use the following code because the third subtopic in the hierarchy changes:

Bedroom/Lighting/LED

Multi-level wildcards

Now, consider a case when you have multiple possible values in the third subtopic hierarchy as well. In that case, using the single-level wildcard won't be enough. Here, we must use the multi-level wildcard, #, which allows you to subscribe to all the subtopic levels. Consider the following code:

Bedroom/Lighting/#

In this case, all the topics that start with `Bedroom/Lightning/` will be subscribed (all the subtopic levels will be included).

Now, the question is, *how does our broker distinguish between clients?* The answer is using *client IDs*. Every MQTT client has a client ID that should be unique according to the protocol rules. In most libraries that we will be using, the system's client ID would be an auto-generated random string to keep the IDs unique.

In the next section, we will discuss MQTT clients in detail.

MQTT clients

MQTT clients were discussed briefly in the first section, but we will go into a little more detail here. I've simplified the definition of an MQTT client as any device that runs an MQTT connection package and connects to an MQTT broker over a local or internet network.

Note that no specific device type is mentioned in the definition. This indicates that a client can be a small microcontroller or microprocessor-based device, or that it can be a full-fledged server. For example, the MQTT client can be a tiny and portable device that connects wirelessly (Wi-Fi) and has a basic MQTT library (for instance, a NodeMCU board). The MQTT client can even be a computer running an MQTT client program for testing purposes. Any device that can use MQTT over TCP/IP can be called an MQTT client.

Next, an MQTT client can either be a *publisher*, *subscriber*, or both. This depends on the application. For example, a computer dashboard would most likely be a subscriber to several MQTT topics as its main task is to show a visual representation of collected data. On the other hand, a *sensor node* will most likely be a publisher who constantly sends the collected sensor data.

We discussed what an MQTT broker is in the previous chapters, so let's move on to the next section, which answers a fascinating question.

How does an MQTT client connect to a broker?

The MQTT broker is another component of an MQTT connection. Its main task is to *manage all the incoming and outgoing messages*. This includes handling all the topics of the network. Moreover, it stores some missed messages if a particular client has opted for the corresponding **Quality of Service (QoS)**. Hence, a single broker can simultaneously handle thousands of clients or more if it is a large-scale implementation.

Now, how exactly does a client connect to a broker either as a subscriber, publisher, or both?

The answer is using something called **MQTT control packets** (in simple terms, this involves exchanging specific information via a network) – a **connect** packet in this case. Whenever a client wishes to connect to a particular MQTT broker, it sends a connect packet to the broker with the necessary attributes. In response, a broker sends a **connect acknowledgment (CONNACK)** packet that contains a status code indicating if the connection was successful and, if not, the reason why it failed. The following diagram visualizes how this process works:

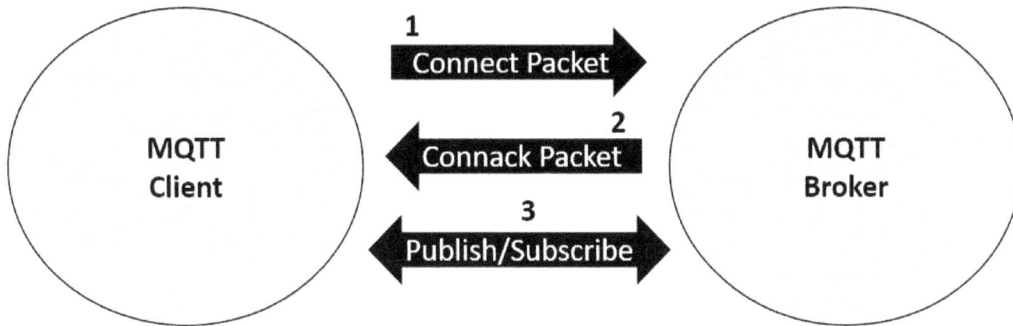

Figure 2.2 – Client-broker connection initiation

> **Important Note**
>
> Please note that in actuality, what happens is that after a client establishes a connection to the broker in a network, the **CONNECT** message is the first message it sends.
>
> Moreover, only one **CONNECT** message can be sent by a single client. If it attempts to send another one, it results in a protocol violation and disconnects from that client.

Now that we know the basics of how MQTT works, we will dive deeper by covering the essential control packets that act as the building blocks of this protocol.

Understanding the MQTT protocol packet structure

MQTT control packets are how the data is managed within an MQTT network. For instance, as discussed in the previous section, when a client wants to connect to a broker, it sends a *connect* packet and, in response, gets a *connack* packet from the broker.

Similarly, when a client wants to publish something on a given topic, it sends a *publish* packet. When a client wants all the data arriving on a particular topic, it achieves that by subscribing to the specific topic using the *subscribe* packet.

In this section, we will provide a detailed discussion of each of the packets. Please note that we will be discussing MQTT v3.1.

Connect packet

The previous section mentions that parameters need to be passed along with a connect packet. The attributes of this message are as follows.

- **Fixed headers**:

The fixed headers only contain the MQTT control packet type (which is the code for **CONNECT**); the remaining length will have a 10-byte variable header, plus the payload.

Bit	7	6	5	4	3	2	1	0
byte 1	MQTT Control Packet type (1)				Reserved			
	0	0	0	1	0	0	0	0
byte 2…	Remaining Length							

Figure 2.3 – Connect packet fixed headers

- **Variable headers**: The variable header for the connect packet has four fields that are in the following order:

  - **Protocol Name** is a UTF-8 encoded string representing the word "MQTT" (in caps).

  - **Protocol Level** is an 8-bit value representing the restriction level of the protocol used by the client. For MQTT v3.1.1, the protocol level value is 0x04. If the broker does not support that protocol, it should send a 0x01 return code in the CONNACK packet. The 0x01 return code indicates that the server's protocol level is unacceptable and unsupported.

  - **Connect Flags** has a byte that contains several parameters specifying the behavior of the MQTT connection. It also indicates the presence or absence of fields in the payload.

  - **Keep-Alive** is a time interval measured in seconds. It is expressed as a 16-bit word; it is the maximum time interval permitted to elapse between the point at which the client finishes transmitting one control packet and the point it starts sending the next.

Please refer to the following table to see what the order of each flag in a byte is:

Bit	7	6	5	4	3	2	1	0
	User Name Flag	Password Flag	Will Retain	Will QoS		Will Flag	Clean Session	Reserved
byte 8	X	X	X	X	X	X	X	0

Figure 2.4 – Connect packet flags

All the flags have a binary value. **Username** and **Password** are two essential attributes that contribute to providing better security to our MQTT network through authentication. The **Username** and **Password** flags indicate whether the payload will contain the username and password.

Will Flag is set to 1. This indicates that if the connect request is accepted, a **Will Message** must be stored on the server and associated with the network connection. The Will Message must be published when the network connection is subsequently closed unless the server has deleted the Will Message on receipt of a **disconnect packet**.

Situations in which the Will Message is published include, but are not limited to, the following:

- An I/O error or network failure has been detected by the server
- The client fails to communicate within the Keep-Alive time
- The client closes the network connection without first sending a disconnect packet
- The server closes the network connection because of a protocol error

There are two bits for **QoS** flags. This is because there are three possible values of QoS, as discussed in *Chapter 1*, *Introduction to the Raspberry Pi and MQTT*. So, 0x00 means that the **Will Flag** is set to 0. If the **Will Flag** is set to 1, the value of `Will QoS` can be 0 (0x00), 1 (0x01), or 2 (0x02). It must not be 3 (0x03). The `Will Retain` flag indicates whether it is to be retained when published. If **Will Flag** is set to 0, then the **Will Retain** flag must be set to 0.

If **Will Flag** is set to 1, then a few scenarios can occur: If **Will Retain** is set to 0, the server must publish the Will Message as a non-retained message. If **Will Retain** is set to 1, the server must publish the Will Message as a retained message.

Finally, the **Clean Session** bit specifies how the session state will be handled.

The client and server can store a session state to enable reliable messaging to continue across a sequence of network connections. This bit is used to control the lifetime of the session state.

If **Clean Session** is set to 0, the server must resume communications with the client based on the current session's state (as identified by the client identifier). If **Clean Session** is set to 1, the client and server must discard any previous session and start a new one.

The following diagram shows a typical **CONNECT** packet:

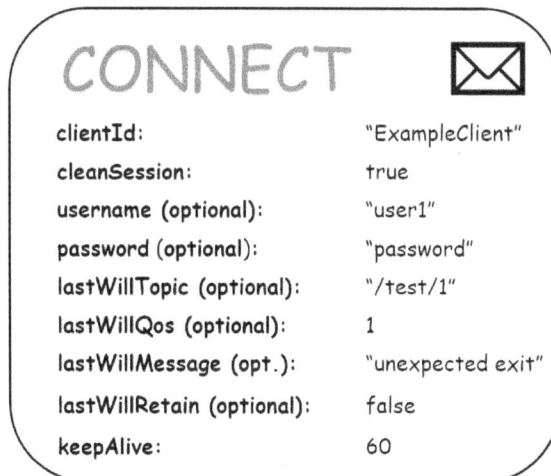

```
CONNECT                        ✉

clientId:                      "ExampleClient"
cleanSession:                  true
username (optional):           "user1"
password (optional):           "password"
lastWillTopic (optional):      "/test/1"
lastWillQos (optional):        1
lastWillMessage (opt.):        "unexpected exit"
lastWillRetain (optional):     false
keepAlive:                     60
```

Figure 2.5 – A typical MQTT connect packet

Next, we will look at the **acknowledgment packet**, which the broker must send due to this connect packet. Then, only the client can publish or subscribe to particular topics.

CONNACK packet

When a broker receives a **CONNECT** message, it must respond with a **CONNACK** message. The CONNACK message contains two entries:

- The session's present flag

- A connect return code

The following diagram shows a typical **CONNACK** packet:

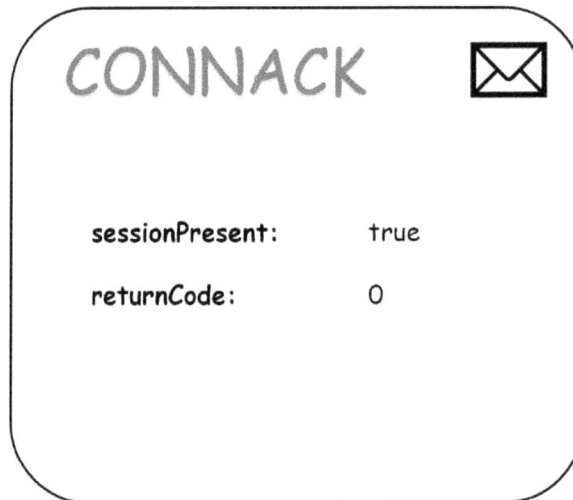

Figure 2.6 – A typical MQTT connack packet

Now, let's understand what the contents in the preceding diagram mean:

- **sessionPresent**: This flag tells the client if there is an existing session available from any previous interactions with that particular client. If a clean session (the **Clean Session** flag in the CONNECT packet is set to `true`), this flag is set to `false`. If the clean session flag is not set, the **ACK** packet will either return the session present flag as `true` if it finds session information for the given client ID or `false` if no session information is available.

- **returnCode**: This flag helps the client determine whether the connection attempt to the MQTT broker was successful or not.

PUBLISH and SUBSCRIBE packets

In a typical MQTT network, there are two basic types of clients:

- **Publishers**
- **Subscribers**

A particular client can be either a subscriber, such as a virtual dashboard, or a publisher, such as a sensor node, or both, which can be a mobile application that can be used to control appliances and also monitor certain sensor values.

The following diagram shows what a potential application could be:

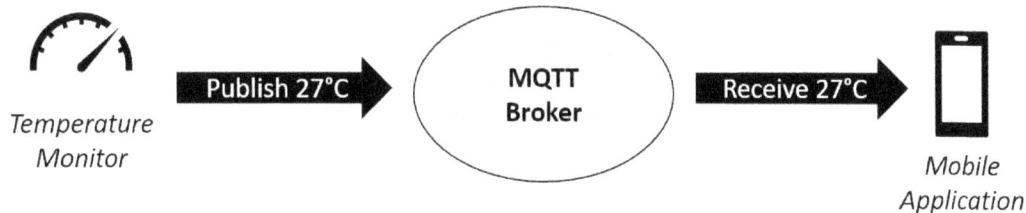

Figure 2.7 – Example of an MQTT application

When it comes to publishers, their primary task is to send the data to the broker on a particular topic. This happens through a **PUBLISH** packet, which is something similar to an **HTTP POST API**.

When the data is collected by a client, it initiates a publish message, which contains several headers and the actual data in the payload of this message.

These headers help the broker keep track of all the publishers and data. In the next section, we will discuss the publish packet in more depth and how a client and the broker use them within an MQTT network.

The same goes for the subscriber. It uses the **SUBSCRIBE** packet to receive all the data or messages on a particular topic. We will also learn about this packet in the next section.

Publish packet

The main purpose of this packet is to send or publish data to a particular channel. The following diagram shows the contents of a general publish packet in MQTT. It has three basic components:

- Variable headers
- Fixed headers
- Payload

A typical MQTT packet is shown in the figure below.

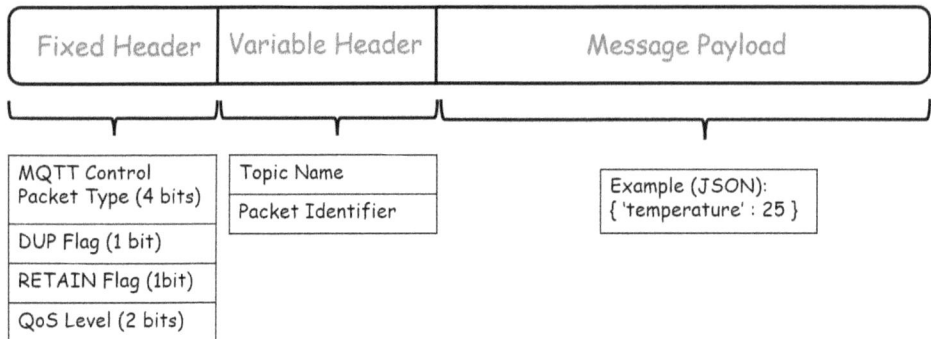

Figure 2.8 – Contents of an MQTT publish packet

Now, let's discuss each of the components of the packet.

Fixed header

This is a 2-byte fixed header that is sent with every publish packet that's initiated either by a client or the server. It contains the following information:

- **MQTT Control Packet Type**: This is a 4-bit value that specifies the type of MQTT packet we are sending. The value for the publish packet is 0x0011.

- The **DUP flag** specifies whether it is the first time this packet is being sent or not (0 means the first time, 1 means resent). Its value is also dependent on the QoS flag. If QoS is 0, this flag is always 0.

- The **Retain flag** specifies whether the message should be retained after being published. This is only available for QoS levels 1 and 2.

- The **QoS Flag** is a 2-bit value that specifies what QoS value the publish packet is sent with. The following table explains this value convention in detail:

Bit 1	Bit 0	QoS Value	Description
0	0	0	At most once
0	1	1	At least once
1	0	2	Exactly once
1	1	-	Reserved

Table 2.1 – Possible values for QoS flags

Variable header

The size of these headers is not fixed as the topic name can be anything and the packet identifier is only specified for packets with a QoS level of 1 or 2. The following list describes the contents of the variable header in detail:

- **Topic Name**: This specifies what topic the message is being published on. An example of a topic is *test-topic/temperature*.

- **Packet Identifier**: This is a unique value that's given to a packet with QoS levels 1 and 2 so that the server can recognize if the message has been sent for the first time or resent.

Payload

The payload contains the actual data of the PUBLISH packet. The size of this data is variable and depends on the type of data being published. Before we look at the next control packet, let's have a look at an example PUBLISH packet to know more about its contents:

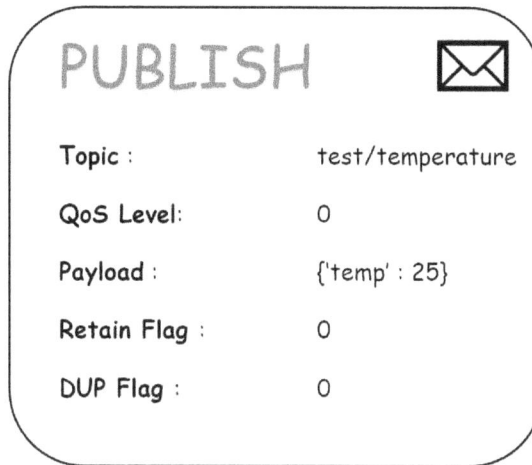

Figure 2.9 – Sample PUBLISH packet

Please note that a response to a PUBLISH packet is required. There are three basic types of responses according to the QoS level, as follows:

- **None**: When the QoS level is 0
- **PUBACK**: When the QoS level is 1
- **PUBREC**: When the QoS level is 2

The subscribe packet will be discussed in detail in the next section. Please note that these are the two main control packets of the MQTT network, so we will limit ourselves to a detailed explanation of these two packets only. You can always check out the original documentation of MQTT for additional details about each control packet:

`http://docs.oasis-open.org/mqtt/mqtt/v3.1.1/os/mqtt-v3.1.1-os.html`

Subscribe packet

The main purpose of this packet is to receive data that's been published on a particular topic. Please keep in mind that this data is sent by the broker and not the publishing client directly. Moreover, a client can subscribe to more than one topic.

The following diagram shows the components of the MQTT SUBSCRIBE packet:

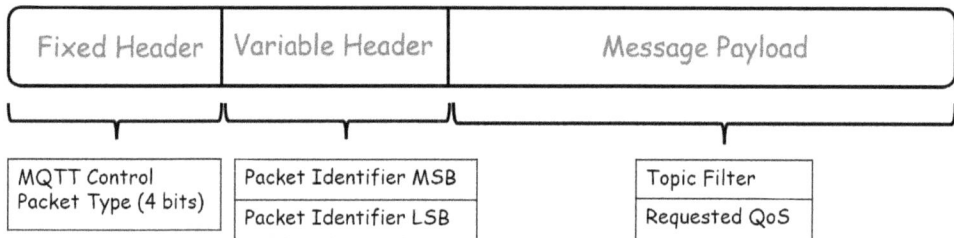

Figure 2.10 – Components of the SUBSCRIBE packet

These components are quite different from the ones in the PUBLISH packet. The reason is that we are not sending any data in the payload, but the topic name that we would like to subscribe to and the QoS level for this subscription.

Now, let's cover each component of this packet in detail.

Fixed header

The fixed header just contains a single attribute the **MQTT Control Packet Type**.

This is a 4-bit value that tells the broker that the packet that's been received is a subscribe packet. The value of this attribute for this particular packet is 0x1000. The rest of the bits of byte 1 are reserved. Byte 2 contains the remaining length of the rest of the packet. This means (length of variable header + length of payload).

Variable header

The variable header also contains two attributes, but they make a single value together: the **most significant bit (MSB)** and the **least significant bit (LSB)** values of the **Packet Identifier**.

Payload

The payload of a SUBSCRIBE packet contains a list of *topic filters* indicating the topics that the client wants to subscribe to, followed by an additional byte with the QoS level for each subscription.

Some important points to note are as follows:

- **Topic Filter** contains the topic names. Hence, they are UTF-8 encoded string values that support wildcard characters.

- The SUBSCRIBE packet must contain at least one topic and QoS value pair. If an empty payload is received, it will be considered a violation of the MQTT protocol and it will throw an error.

- The **Requested QoS** byte has 6 reserved bits and only the last 2 bits are used to specify the QoS level for a particular topic.

A sample SUBSCRIBE packet looks something like this:

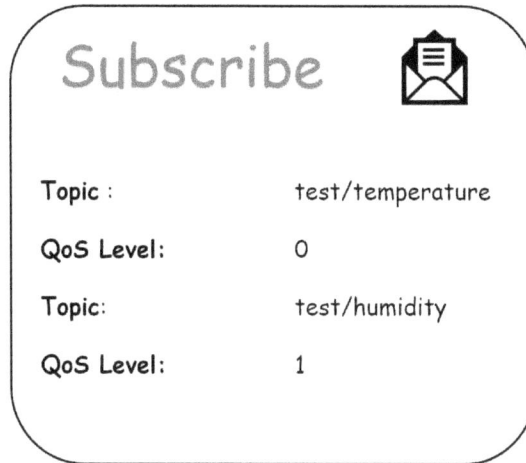

Figure 2.11 – Sample SUBSCRIBE packet

After a SUBSCRIBE packet has been sent to the broker, according to protocol, it must respond with a **SUBACK** packet. Here are some important points to know about the behavior of the SUBACK packet:

- The SUBACK packet must have the same packet identifier as the SUBSCRIBE packet that it is acknowledging.

- The server is permitted to start sending PUBLISH packets that match the subscription before the server sends the SUBACK packet.

- The SUBACK packet sent by the server to the client must contain a return code for each *Topic Filter/QoS* pair. This return code must either show the maximum QoS that was granted for that subscription or indicate that the subscription failed.

This is the end of the MQTT control packets subsection. Now comes the most important section of this chapter, which is seeing how the process works.

Practical demonstration of MQTT in action

In this section, we will look at how the communication protocol works under the hood. For this, we will use an additional software called **Wireshark**, which lets us capture all the incoming and outgoing MQTT packets from our broker hosted on the Raspberry Pi.

Here, we will look at how a subscriber or a publisher connects to and sends the message through our MQTT broker. So, let's get started:

1.   The first step is to install Wireshark on our Raspberry Pi. To do so, open your Terminal and type in the following command:

     ```
     sudo apt install wireshark
     ```

     It will ask you to confirm that you wish to install the package. Just type Y and press *Enter*. This will start the installation process. When this happens, you will see a configuration screen appear in front of you, as shown in the following screenshot. It will ask if you want to allow any user to have maximum privileges for the Wireshark application. I suggest that you choose **No** for this part as it is a very bad practice to do so:

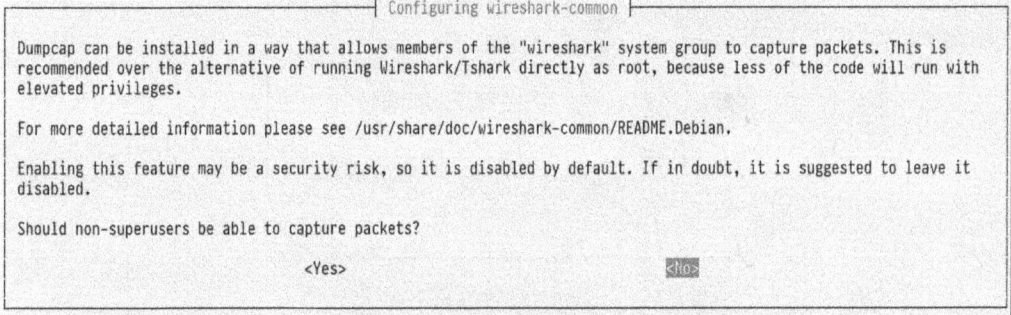

```
───────────────────┤ Configuring wireshark-common ├───────────────────

  Dumpcap can be installed in a way that allows members of the "wireshark" system group to capture packets. This is
  recommended over the alternative of running Wireshark/Tshark directly as root, because less of the code will run with
  elevated privileges.

  For more detailed information please see /usr/share/doc/wireshark-common/README.Debian.

  Enabling this feature may be a security risk, so it is disabled by default. If in doubt, it is suggested to leave it
  disabled.

  Should non-superusers be able to capture packets?

                      <Yes>                                        <No>
```

Figure 2.12 – Wireshark configuration screen

     This will complete the installation process.

2.   Next, we will add a new user group specifically for this application so that we can add the users for which access to this application is allowed. I will name the group wireshark and add the pi user (please use the hostname) to this group using the following commands:

     ```
     sudo groupadd wireshark
     sudo usermod -a -G wireshark pi
     ```

3. Next, we will need to execute a series of commands so that we can grant the necessary permissions to our user group for all the required files. Just follow the given instructions to achieve this. What these commands do is grant read and write execution of the dumpcap file to our group (where all packet captures are stored) and configure certain capabilities:

```
sudo chgrp wireshark /usr/bin/dumpcap
sudo chmod 750 /usr/bin/dumpcap
ls -al /usr/bin/dumpcap
sudo setcap cap_net_raw,cap_net_admin=eip /usr/bin/
dumpcap
sudo getcap /usr/bin/dumpcap
```

4. Now, reboot your Raspberry Pi. With that, you have finished configuring Wireshark. You should be able to run this application by simply entering wireshark in your Terminal.

5. Once you execute the wireshark command on your Terminal, a new window will open with a heading stating The Wireshark Network Analyzer.

    It will show all the available network interfaces that can be analyzed. The following screenshot shows what the home screen looks like:

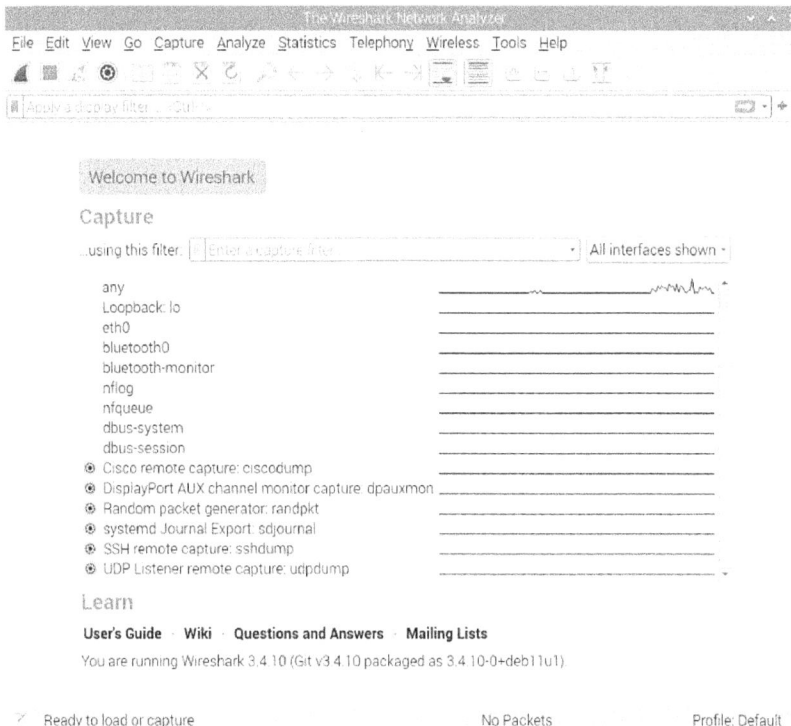

Figure 2.13 - Wireshark home screen

6. Next, select the wlan0 option as that is the option where we can capture all the packet captures that are sent or received wirelessly over our Wi-Fi network.

7. With that, we are all set up, but there are still things that will greatly help us. The first thing you will want to do is add a display filter as if you are using your Pi over VNC. You will see several packets being captured every second, which makes it very difficult to look for the captures that are of interest to us. For that, simply type mqtt in the **Apply a display filter** text box and press *Enter*.

You can use the following screenshot as a reference:

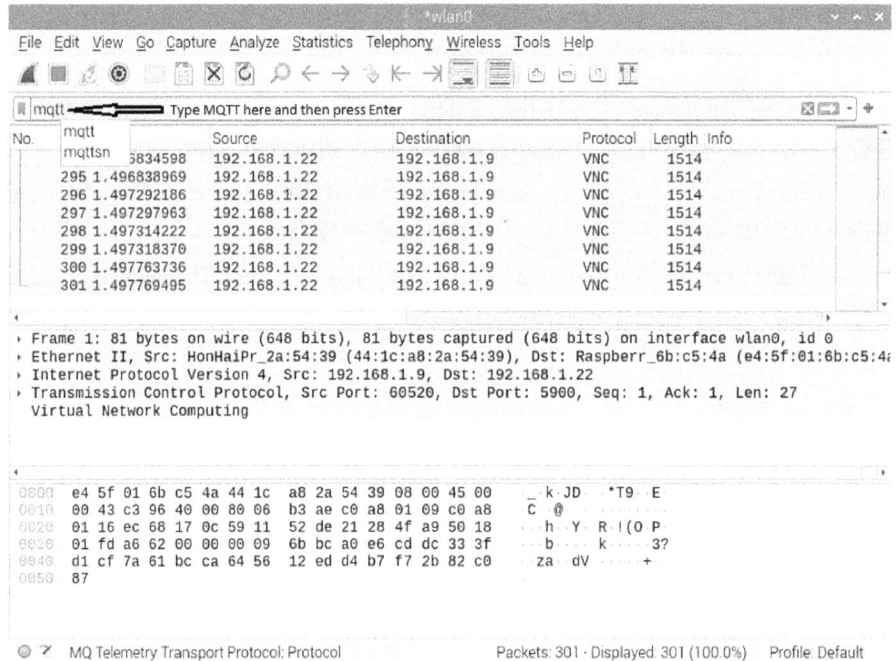

Figure 2.14 – Applying a display filter in Wireshark

8. Upon doing this, an empty screen will appear with no captures as we have not connected any clients to our broker. We will need to connect with remote clients so that data packets can be transferred via the wireless network. If you have a Linux or Mac system, this is very easy to do. Just type in the following command:

```
sudo apt install mosquitto-clients
```

This will let you access the mosquitto_pub and mosquitto_sub commands from your Terminal. For Windows, the process is a little different. First, head to the official Mosquitto website's download page: https://mosquitto.org/download/.

Just download the Windows `.exe` file for your Windows system (32 or 64-bit) and run the installer. Just keep pressing **Next** and then install all the required components. It will take some time, so please be patient. Once it is complete, you will be able to use the `mosquitto_pub` and `mosquitto_sub` commands on your Windows system.

Please note that to run these commands, you will have to navigate to the `mosquitto` directory, whose path you will have mentioned during installation. If you kept the default path, you would find the *mosquitto* folder under `C:\Program Files\mosquitto`.

9. To test if the installation was successful, we will open two Command Prompt or Terminal windows on our main system (according to the OS you have) and type in the following commands, respectively:

```
mosquitto_sub -h <ip of your pi> -v -t sensors/
temperature
mosquitto_pub -h <ip of your pi> -t sensors/temperature
-p "27"
```

Upon entering these commands, the number 27 will appear on the window where you ran the `mosquitto_sub` command when you execute the `mosquitto_pub` command.

If this was successful, then you have successfully set up mosquitto on your main system. Now, let's look at how MQTT works by looking at the packets that are generated after executing each command.

10. To do so, open the desktop of your Raspberry Pi. You should have the Wireshark window open with the MQTT display filter already applied. To start the packet captures, just press the button marked in the following screenshot:

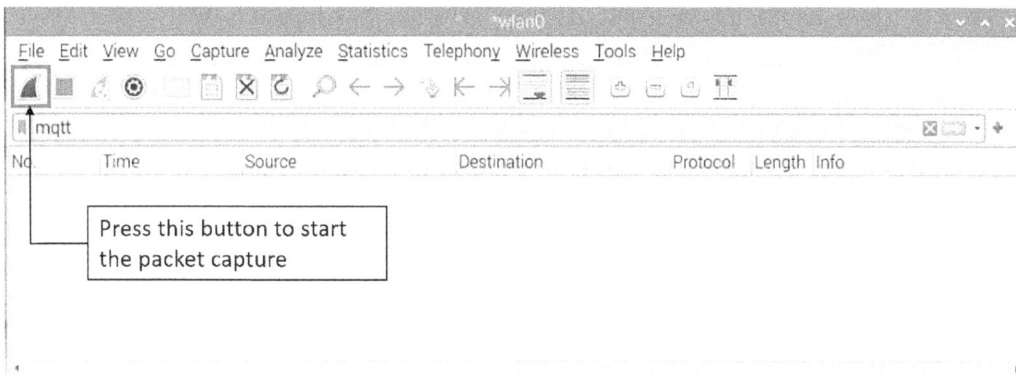

Figure 2.15 – Press the marked button to start the packet captures

11. Now, the application will capture and show all the MQTT packets. Next, go to your main system and repeat *Step 6*, executing the two commands mentioned. You will see the output shown in the following screenshot on the Wireshark application:

No.	Time	Source	Destination	Protocol	Length Info
157	7.544469...	192.168.1.9	192.168.1.22	MQTT	68 Connect Command
159	7.545236...	192.168.1.22	192.168.1.9	MQTT	58 Connect Ack
160	7.548200...	192.168.1.9	192.168.1.22	MQTT	80 Subscribe Request (id=1) [sensors/temperature]
162	7.548508...	192.168.1.22	192.168.1.9	MQTT	59 Subscribe Ack (id=1)
251	9.716409...	192.168.1.9	192.168.1.22	MQTT	68 Connect Command
253	9.717046...	192.168.1.22	192.168.1.9	MQTT	58 Connect Ack
254	9.719985...	192.168.1.9	192.168.1.22	MQTT	79 Publish Message [sensors/temperature]
256	9.720209...	192.168.1.9	192.168.1.22	MQTT	56 Disconnect Req
257	9.720243...	192.168.1.22	192.168.1.9	MQTT	79 Publish Message [sensors/temperature]

Figure 2.16 – Packet captures for the experiment

12. Next, we will dissect the packet captures to understand how the process works. The software will assist us in doing so. We will break down the packets into two parts: the *subscribe packets* and the *publish packets*. Starting with the subscribe packets, please refer to the following screenshot, which shows all the corresponding packets:

```
MQTT        68 Connect Command
MQTT        58 Connect Ack
MQTT        80 Subscribe Request (id=1) [sensors/temperature]
MQTT        59 Subscribe Ack (id=1)
```

Figure 2.17 – MQTT subscribe packets

Here, we can see that four packets are listed. The first two are the **Connect** and **Connack** packets, which help the client (in this case, our main system) establish a connection with our broker. We will start with these first.

CONNECT packet

This packet is sent from the client to the broker requesting a connection. To see more information about the contents of this packet, just double-click on the first row, which states **Connect Command**; this will open a new window where you will be able to see a breakdown of the packet. Expand the MQTT part and you will be able to see all the contents in detail. The following screenshot shows the relevant MQTT contents that are a part of this packet:

```
▾ MQ Telemetry Transport Protocol, Connect Command
  ▸ Header Flags: 0x10, Message Type: Connect Command
    Msg Len: 12
    Protocol Name Length: 4
    Protocol Name: MQTT
    Version: MQTT v3.1.1 (4)
  ▾ Connect Flags: 0x02, QoS Level: At most once delivery (Fire and Forget), Clean Session Flag
      0... .... = User Name Flag: Not set
      .0.. .... = Password Flag: Not set
      ..0. .... = Will Retain: Not set
      ...0 0... = QoS Level: At most once delivery (Fire and Forget) (0)
      .... .0.. = Will Flag: Not set
      .... ..1. = Clean Session Flag: Set
      .... ...0 = (Reserved): Not set
    Keep Alive: 60
    Client ID Length: 0
    Client ID:
```

Figure 2.18 – MQTT CONNECT packet dissection

Now, let's dive deeper into the meaning of the preceding code. We'll cover this in a similar format to when we covered the CONNECT packet.

Fixed header

The fixed header is a single byte that specifies the control packet type. In the case of the CONNECT packet, its value is 0001 0000 (10 in hex format), which is the value of the header flags. It even displays the interpretation, which is a plus point for us.

Variable header

Next are the variable headers, which contain the following information:

- Protocol Name

- Protocol Level

- Connect Flags

- Keep-Alive

As shown in the preceding screenshot, all this information is listed. **Protocol Name** is *MQTT* (the hex value is 4d 51 54 54), while **Protocol Level** is the version of MQTT our broker is using – that is, MQTT v3.1.1. For that, the protocol version of *4* has been given. This is one of the reasons why MQTT may have skipped MQTT v4 and went straight to v5 – to avoid any confusion in protocol levels.

Then, comes **Connect Flags**, which is a single byte that carries a lot of information. We discussed the significance of each flag when we discussed this packet, so you can just refer to the screenshot for the values. The outcome is an *unprotected request* (no username and password) with *no retention, QoS value of 0, no At Will behavior*, and a *new session is started* (the clean session flag is set to 1).

Finally, the next two bytes will give you the duration for which this connection will remain active, which is 60 seconds by default. After that, a ping request will be sent by the client; if it receives a ping acknowledgment, the connection persists. The following diagram shows the ping packets, along with their contents, that have been captured by Wireshark:

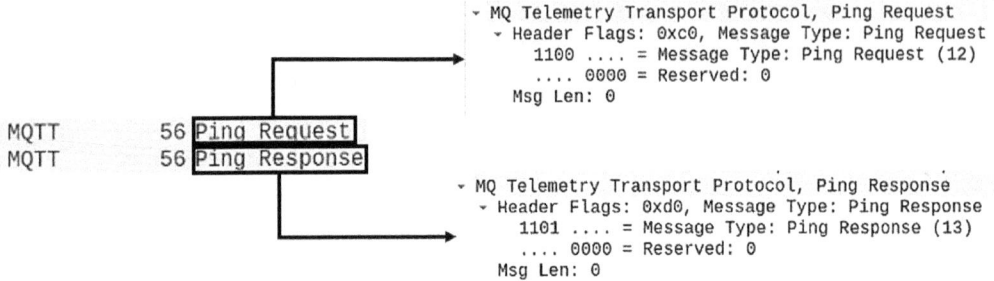

```
                                        ⌄ MQ Telemetry Transport Protocol, Ping Request
                                          ⌄ Header Flags: 0xc0, Message Type: Ping Request
                                              1100 .... = Message Type: Ping Request (12)
                                              .... 0000 = Reserved: 0
                                            Msg Len: 0

    MQTT        56 Ping Request
    MQTT        56 Ping Response

                                        ⌄ MQ Telemetry Transport Protocol, Ping Response
                                          ⌄ Header Flags: 0xd0, Message Type: Ping Response
                                              1101 .... = Message Type: Ping Response (13)
                                              .... 0000 = Reserved: 0
                                            Msg Len: 0
```

Figure 2.19 – Ping packets dissected for ease of understanding

Payload

This is yet another variable-length component of this packet. It contains the username and password, provided their corresponding control flags have been set and the client ID length and the actual client ID are also a part of the variable component. Only the last two components are present for our test case, as shown in *Figure 2.18* (the last two lines).

CONNACK packet

This is the acknowledgment package that's sent by our broker to the client. It consists of an acknowledgment message that says whether the connection has been established or not.

As we know, this packet has three components, as follows:

```
⌄ MQ Telemetry Transport Protocol, Connect Ack
  ⌄ Header Flags: 0x20, Message Type: Connect Ack
      0010 .... = Message Type: Connect Ack (2)
      .... 0000 = Reserved: 0
    Msg Len: 2
  ⌄ Acknowledge Flags: 0x00
      0000 000. = Reserved: Not set
      .... ...0 = Session Present: Not set
    Return Code: Connection Accepted (0)
```

Figure 2.20 – MQTT CONNACK packet dissection

Fixed header

This contains the same 2 bytes that tell you the type of the MQTT control packet. In this case, the value will be 20 (0010 0000 – byte value).

Acknowledgment flags

These 2 bytes only give you the value of the session's present flag, which states whether an existing session is going on for that client. It is not set in our case.

Return code

Finally, we have the return code, which states whether a connection has been established with the broker. A 00 2-byte value suggests that the connection was accepted by the broker.

SUBSCRIBE and SUBACK packets

Once the connection has been established, the client will send a packet requesting to be a subscriber, along with all the relevant information.

Refer to the following screenshot to see what the main components of this packet are:

```
▼ MQ Telemetry Transport Protocol, Subscribe Request
  ▼ Header Flags: 0x82, Message Type: Subscribe Request
      1000 .... = Message Type: Subscribe Request (8)
      .... 0010 = Reserved: 2
    Msg Len: 24
    Message Identifier: 1
    Topic Length: 19
    Topic: sensors/temperature
    Requested QoS: At most once delivery (Fire and Forget) (0)
```

Figure 2.21 – MQTT SUBSCRIBE packet dissection

As you can see, there is a lot of information in this packet. Let's look at this information in more detail.

Fixed header

The header value for the SUBSCRIBE packet is 80 (1000 0000 in binary format). This is a compulsory inclusion in each MQTT packet so that the broker and clients know the type of control packet that's being used.

Variable headers

Next are the variable headers, which contain the message's length (the length of the rest of the message). Including the message identifier and payload, we have 24 bytes. The message identifier is a 2-byte value with a value of 0001 (hex format).

Payload

Finally, the payload contains the topic's name, its length, and the requested QoS by the client.

Once the broker receives this SUBSCRIBE request packet, it will reply to the client with a `Subscribe Ack` packet, which will contain the same information as the CONNACK packet but with some exceptions (the message length and message identifier bytes). The following screenshot shows what data this packet conveys:

```
▾ MQ Telemetry Transport Protocol, Subscribe Ack
  ▾ Header Flags: 0x90, Message Type: Subscribe Ack
      1001 .... = Message Type: Subscribe Ack (9)
      .... 0000 = Reserved: 0
    Msg Len: 3
    Message Identifier: 1
    Granted QoS: At most once delivery (Fire and Forget) (0)
```

Figure 2.22 – Subscribe Ack packet dissection

With that, you know how a client subscribes to a particular topic. Now, let's see how a client publishes a message on a specific topic:

1. First, we will analyze all the packets that are sent after executing the `publish` command. As we have already executed the command, you should have the captured packets listed in Wireshark after the SUBSCRIBE packets.

    The following screenshot shows the packets that are of interest to us:

    ```
    MQTT          68 Connect Command
    MQTT          58 Connect Ack
    MQTT          79 Publish Message [sensors/temperature]
    MQTT          56 Disconnect Req
    ```

    Figure 2.23 – MQTT publish command packet capture

The CONNECT and CONNACK packets will be the same as they were for the SUBSCRIBE packet. They will contain the same information as this is a common process that happens whenever a client, be it a publisher or subscriber, tries to establish a connection with the MQTT broker, which is the Raspberry Pi in our case.

So, we won't explain these packets as they have already been covered. Just for reference, please refer to the following screenshot, which shows the contents of both the **CONNECT** and **CONNACK** packets for the `publish` command:

CONNECT Packet

```
~ MQ Telemetry Transport Protocol, Connect Command
  ~ Header Flags: 0x10, Message Type: Connect Command
      0001 .... = Message Type: Connect Command (1)
      .... 0000 = Reserved: 0
    Msg Len: 12
    Protocol Name Length: 4
    Protocol Name: MQTT
    Version: MQTT v3.1.1 (4)
  ▸ Connect Flags: 0x02, QoS Level: At most once delivery (Fire and Forget), Clean Session Flag
    Keep Alive: 60
    Client ID Length: 0
    Client ID:
```

CONNACK Packet

```
~ MQ Telemetry Transport Protocol, Connect Ack
  ~ Header Flags: 0x20, Message Type: Connect Ack
      0010 .... = Message Type: Connect Ack (2)
      .... 0000 = Reserved: 0
    Msg Len: 2
  ~ Acknowledge Flags: 0x00
      0000 000. = Reserved: Not set
      .... ...0 = Session Present: Not set
    Return Code: Connection Accepted (0)
```

Figure 2.24 – The CONNECT and CONNACK packets for the publish command

We will discuss the PUBLISH packet now, which is the packet that commands the broker to publish a message on a particular topic. We discussed the structure of this packet in the previous section. Before we look at each component of the packet, please refer to the following screenshot, which shows what the components of this packet are:

```
MQ Telemetry Transport Protocol, Publish Message
~ Header Flags: 0x30, Message Type: Publish Message
    0011 .... = Message Type: Publish Message (3)
    .... 0... = DUP Flag: Not set
    .... .00. = QoS Level: At most once delivery (Fire and Forget) (0)
    .... ...0 = Retain: Not set
  Msg Len: 23
  Topic Length: 19
  Topic: sensors/temperature
  Message: 3237
```

Figure 2.25 – MQTT PUBLISH packet breakdown

As you can see, there are some changes compared to the previous packets that we have covered. Let's explore them together.

Fixed header

The most noticeable change is that in the header flags or the fixed header, along with the *MQTT control packet type*, whose value is 0x30 (the code for the PUBLISH packet), we have three additional flags. **DUP Flag** indicates if this is the first time this packet is being sent by this client. Its value depends on the QoS value. For a QoS value of 0, this flag will always have a value of 0, as shown in the preceding screenshot. **QoS Level** is shown by 2 bits. Currently, we are operating on a value of 0, which means it will execute this command once and that's all. No reply is required. Finally, we have the **Retain** flag, which tells the broker if they have to retain the message after sending it. Please note that it is only for QoS levels 1 and 2 and hence it is set to 0 now.

Variable header

The variable header consists of the various components, as shown in the preceding screenshot – it conveys the *message and topic length* (in bytes) and also contains the *topic name*.

Payload

This is the last component of this packet, and it contains the *message content* that we want to publish on the aforementioned MQTT topic name (that is, **sensors/temperature**). In our case, the message we sent was 27. Now, if you refer to the breakdown, you will see that the value of the message was printed as 3237. Why is that?

The answer is the encoding of this message. All the messages follow UTF-8 encoding, and according to that, the value of hex code 32 means digit 2, while the value of hex code 37 means digit 7. Let's refer to the following UTF-8 encoding table:

1	31	DIGIT ONE
2	32	DIGIT TWO
3	33	DIGIT THREE
4	34	DIGIT FOUR
5	35	DIGIT FIVE
6	36	DIGIT SIX
7	37	DIGIT SEVEN
8	38	DIGIT EIGHT

Figure 2.26 – Hex code to UTF 8 value conversion

Finally, once the publish packet has been sent, the client will send the **Disconnect-Request** packet, which means that all the required transactions are over, so it would like to disconnect from the broker.

It sends a packet that has a similar structure to an acknowledgment packet. Its content can be seen in the following screenshot:

```
MQ Telemetry Transport Protocol, Disconnect Req
  ▾ Header Flags: 0xe0, Message Type: Disconnect Req
      1110 .... = Message Type: Disconnect Req (14)
      .... 0000 = Reserved: 0
    Msg Len: 0
```

Figure 2.27 – MQTT Disconnect-Request packet breakdown

This topic completes this chapter. Now, let's summarize what we covered.

Summary

We covered a lot of content about MQTT in this chapter. This will help you a lot in the upcoming chapters, especially the two projects that we will be creating since you will be aware of how devices are communicating with the Raspberry Pi's MQTT broker.

In this chapter, we started by introducing the communication protocol, and we explored the main components or building blocks of MQTT. Next, we understood how the protocol works and covered some important control packets, which are the building blocks of MQTT.

Finally, we explored a practical scenario and understood how the actual communication happens under the hood by using a packet capturing software called Wireshark on our Raspberry Pi.

In the next chapter, we will cover some popular development boards manufactured by **Espressif** – the **ESP8266**-based **NodeMCU development board** and the **ESP32 development board**. We will demonstrate how to use these boards as an MQTT client and communicate with our Pi.

3
Introduction to ESP Development Boards

In this book, we will be using two types of ESP-based development boards manufactured by **Espressif Systems (688018.SH)**. Espressif Systems is a public multinational, fabless semiconductor company established in 2008. They mainly develop state-of-the-art Wi-Fi and Bluetooth-based IoT development boards and SoCs. Their popular products include the ESP8266 (the chipset powering the popularly known NodeMCU), ESP32, ESP32-S, and ESP32-C series of chips, modules, and development boards.

This chapter introduces you to two of the most popular development boards: the ESP8266 based NodeMCU and the ESP32 development board. The chapter is divided into 3 main sections:

- **ESP8266**-based NodeMCU development board
- **ESP32**-based development board
- **Mini-Project 1**: NodeMCU as an MQTT client

This chapter will be divided into three main sections. The first two sections will provide details about these development boards, discussing each point listed as follows:

- Technical specifications of the development board
- The pinout diagram and GPIO configuration
- Software setup to program these boards using the Arduino IDE

Finally, we will create our very first mini-project, where we will set up a NodeMCU development board as an MQTT client. This will connect to our Raspberry Pi MQTT broker and control its onboard LED using the terminal of our home computer. *Exciting*, right?

Let's waste no more time and dive in!

ESP8266-based NodeMCU development board

NodeMCU is an open source development board that is designed to prototype IoT applications. The development board equips the ESP-12E module, which contains an ESP8266 chip. This chip has a *Tensilica Xtensa® 32-bit LX106 RISC microprocessor* that operates at an 80 to 160 MHz-adjustable clock frequency and supports RTOS.

The board can be programmed using two languages, as follows:

- Embedded C (using the popular Arduino IDE)

- Lua Programming Language

We will learn how to program NodeMCU through the Arduino IDE later in this chapter.

First, let's look at the actual development board. The following is a diagram of the NodeMCU board with the important peripherals of the board labeled accordingly:

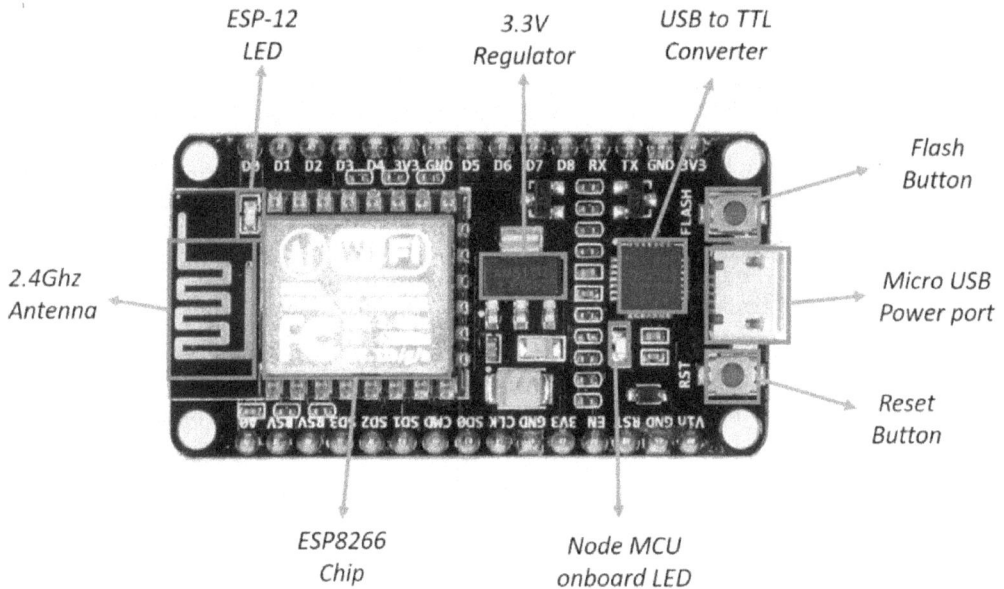

Figure 3.1 – A NodeMCU development board

Next, we will look at the technical specifications for this development board.

Technical specifications

The development kit that's based on ESP8266 integrates GPIO, PWM, IIC, and the 1-Wire and ADC all-in-one board. You can power your development in the fastest way by combining this with NodeMCU firmware. The technical specifications for the NodeMCU development board are as follows:

- **Wi-Fi Module**: An ESP-12E (32-bit) module that's similar to the ESP-12 module but with six extra GPIOs that support the *802.11 b/g/n* Wi-Fi protocol.

- **Power Source**: A micro USB port for power, programming, and debugging.

- **Headers**: A 2 x 2.54 mm 15-pin header with access to GPIOs, SPI, UART, ADC, and power pins. We will discuss the GPIO pinout in detail in the next section.

- **Power Rating**: The required power to power the board is 2.6 to 3.3 V with a current of 250 mA. The USB port provides 5 V, which is regulated to 3.3 V by an on-chip AMS1117 voltage regulator board.

- **Dimensions**: 49 x 24.5 x 13 mm.

- **Flash Memory**: 4 MB.

- **SRAM**: 64 KB.

- **ADC Pins**: It has 1 ADC pin, which has a voltage range of 0 - 3.3 V.

- **Digital Pins**: It has 11 digital I/O pins.

- **Miscellaneous**: Reset and Flash buttons.

- **Temperature Range**: The company has rated the temperature range for the product to be between -40 degrees to 125 degrees Celsius.

- **Price**: The NodeMCU board retails in the price range of $2 to $5.

Next, we will look at the pins that the development board boasts and the configuration and functionality of each pin.

NodeMCU GPIO pinout and pin configurations

The ESP8266 NodeMCU board has a total of 30 pins that can be used to connect it to any peripherals or development boards.

The following diagram shows the GPIO pinout for the NodeMCU development board:

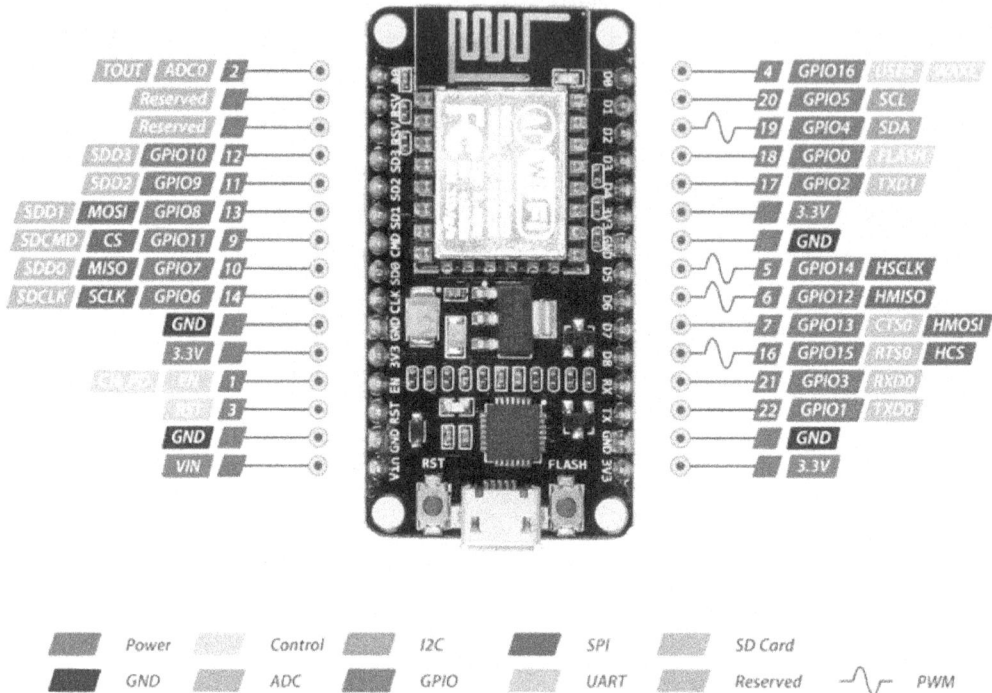

Figure 3.2 – NodeMCU detailed pinout diagram

The preceding diagram seems considerably complex. So, let's cover it in parts to make understanding it easy. The first thing to notice is that there are several pins with more than one box. What this means is that these pins have more than one possible functionality, but we can only use one at a given moment. Such pins are called **multiplexed pins**. We can select our preferred mode for that pin using our code. We will see how this works practically later in this book.

Now, let's cover the pins by using the legend provided in the preceding diagram:

- Firstly, the **power pins** are marked in red. As the name suggests, these pins are used to supply power to external components. There are four power pins, three of which output 3.3 V (the output of the onboard 3.3 voltage regulator), and one VIN pin, which outputs the raw input voltage when the board is powered through the on-chip micro USB port. Alternatively, it can be used to power the board through an external battery.

- There are 4 GND or **ground pins** on the board as well.

- The **control pins** are used to control the ESP8266 chip. There are three such pins of importance on the board: the **Enable (EN)**, **Reset (RST)**, and **Wake (WAKE)** pins. They have different functionalities, as follows:

  - The **Enable pin** is used to enable or disable the ESP8266 chip. Supplying a HIGH or 3.3 V signal will enable the ESP8266 chip, while a LOW or GND signal will operate the ESP chip on minimum or low power mode.

  - The **Reset pin**, as its name suggests, is used to reset the ESP8266 chip.

  - The **Wake pin** helps us wake the ESP8266 chip from deep sleep (one of the low-power modes).

- The **ADC pins** are a part of the 10-bit ADC that the NodeMCU board possesses. The maximum voltage for this ADC is 1 V (0 to 1 V). There are two pins on the board: the ADC0 and TOUT pins.

- The **UART pins** provide a serial communication interface. The NodeMCU board has two such interfaces: UART0 and UART1. The maximum communication speed is 4.5 Mbps. The UART0 interface consists of TXD0, RXD0, CTS0, and RST0 pins to enable both transmission and reception of signals. On the other hand, the UART1 interface just has a TXD1 pin, so it only supports transmission capabilities. One example application for such an interface could be logging.

- The **SPI pins** provide an SPI communication interface. The NodeMCU board has two SPI interfaces: SPI and HSPI. There are eight SPI pins on the board, four for each interface.

- The ESP8266 features an **SDIO interface** as well, which can be used to directly interface SD cards. 4-bit 25 MHz SDIO v1.1 and 4-bit 50 MHz SDIO v2.0 are supported. There are six pins for this interface on the board.

- There are 17 **GPIO pins** on the NodeMCU board, which can be used to connect peripherals and other boards. It can be assigned various functions as the majority of these pins are multiplexed. You can pull the pin up or down internally through software commands. It can act as an input and can be set to trigger when it receives a certain signal.

- Additionally, the board supports the **I2C communication interface**. Both master and slave functionality is supported, with a maximum clock frequency of 100 kHz. There are two pins for this interface on the board: the SDA (data) and SCL (clock) pins.

- Finally, there are two reserved pins on the board. Additionally, some GPIO pins have a wave-like symbol beside them, which signifies that they are **PWM (Pulse Width Modulation) pins**. The PWM frequency range for this board is 100 Hz to 1 kHz.

This concludes this section. Now that we know a lot about the NodeMCU development board (both its hardware and its GPIO pinout), let's learn how to set up the Arduino IDE so that we can program this board through that software. We can even set up our Raspberry Pi to program this board, but for the sake of simplicity, we will stick to the PC setup.

Arduino IDE setup for the NodeMCU development board

There are two ways to program the board: we can either use **Arduino-based C programming** or the **Lua programming language**.

We will stick to the Arduino IDE setup for this tutorial as it is easier to follow and much more reliable. Follow these steps to successfully run your first program on the NodeMCU board:

1. First, go to https://www.arduino.cc/en/software and download the latest stable version of Arduino IDE for your computer. This can be seen in the following screenshot:

Figure 3.3 – Choose the download option for your computer's OS

2. Once the installer (executable file) has been downloaded, just install the IDE on your computer. To do so, just follow the onscreen steps and keep pressing the **Next** button.

> **Note**
>
> For Windows 8.1 or higher, you can just download the Arduino app from Microsoft Store. If you downloaded a ZIP file, just extract the contents, copy the extracted Arduino folder to the desired destination, and launch the application.

3. Once the IDE is successfully installed on your computer, open it through the app icon or from the start or search menu, depending on the operating system you are using.

Although the development environment has been installed on your computer, it does not support the programming of ESP-based boards out of the box. We need some additional setup for that. For this, you must go to **File | Preferences**, as shown in the following screenshot:

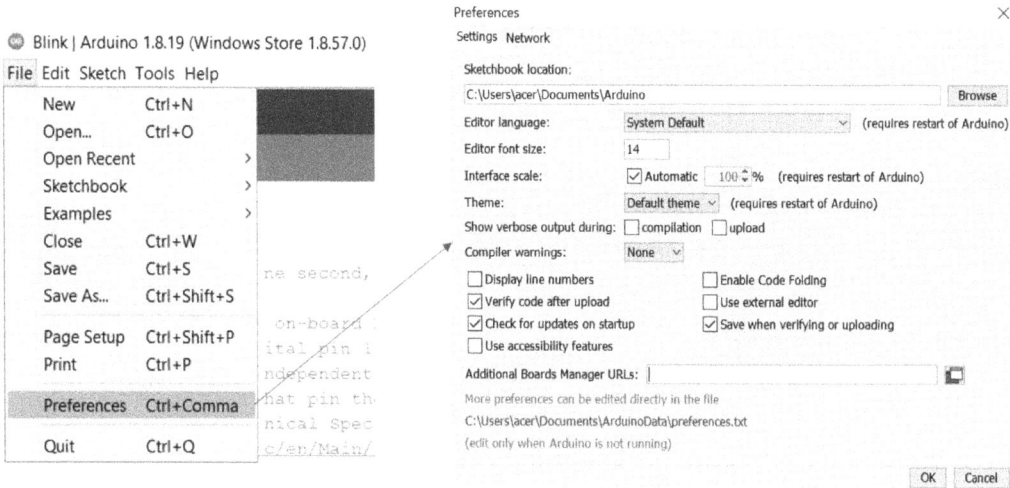

Figure 3.4 – Arduino IDE – Preferences

4. In the **Additional Board Manager URLs** area, paste the following URL:

```
http://arduino.esp8266.com/stable/package_esp8266com_index.
json
```

5. Now, press **OK**.

This will allow you to download the necessary package files for ESP8266 development from **Boards Manager**.

6. The next step is to download all the packages and files needed by the Arduino IDE. To do that, go to **Tools | Board: "Arduino Uno" | Boards Manager**.

7. In the search bar, search for `esp8266`. You will see find an option by that name, thanks to the URL we entered:

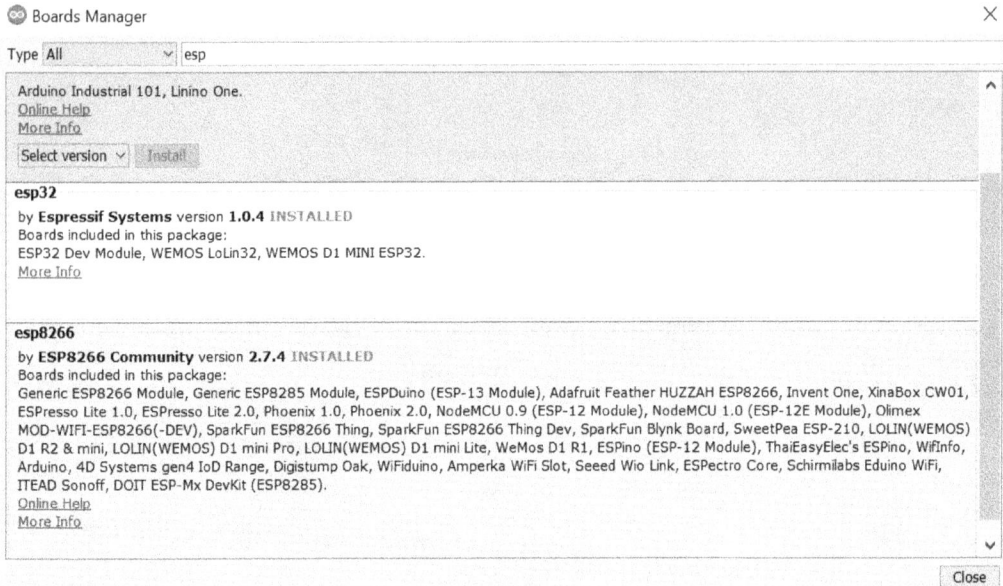

Figure 3.5 – Arduino IDE Boards Manager

8. Install the **esp8266** package (the latest version) by pressing the **Install** button. It will take around 4 to 5 minutes to download and install all the requirements.

Now, you are ready to program your ESP8266-based development boards!

Let's verify this by uploading the **LED Blink** code onto our NodeMCU board:

1. Open the example Blink code by going to **File | Examples | 01. Basics | Blink**.

    This will open the LED Blink example code in a new window. The code is very simple, as shown here:

    ```
    void setup ()
    {
      pinMode (LED_BUILTIN, OUTPUT);
    }


    void loop ()
    {
      digitalWrite (LED_BUILTIN, HIGH);
    ```

```
    delay (1000);

    digitalWrite (LED_BUILTIN, LOW);
    delay (1000);
}
```

2. Now, we need to select the **Node MCU dev** board from the **Boards** menu so that the IDE recognizes our board. To do so, go to **Tools** | **Board: 'x'** | **ESP8266 Boards** | **NodeMCU 1.0**.

    The next step involves connecting the board and selecting the appropriate COM port. This specifies which USB port your NodeMCU board is connected to.

> **Important Note**
>
> If you are using an older version of the Arduino IDE, you may see a single list of boards instead of the **ESP8266 Boards** section. Just scroll down the list to find the board mentioned previously and click on it.

3. Connect your NodeMCU board to your PC using a USB to micro USB cable. Once you've done this, your computer may install some drivers if you are connecting the device for the first time, but this process does not usually occur in newer systems. Make sure that any other USB devices connected to the computer are disconnected.

4. Now, the COM port for this device needs to be selected. This can be done by going to **Tools** | **Port**.

    You should see **COMxx** for a Windows PC and **dev/ttyUSBx** for a Linux or Mac PC. Just select the port by pressing on it.

5. Finally, press the **Upload** button, which can be found beside the tick icon on the menu on the top right-hand side of the Arduino IDE window. This can be seen in the following screenshot:

Figure 3.6 – The Upload button

6. If everything works fine, the bottom window will show that the program has been successfully uploaded onto your development board and the on-board LED will start blinking in 1-second intervals.

This completes our introduction to the NodeMCU development board. The next section will introduce you to another popular development board from Espressif, the **ESP32 development board**. It has several additional features, such as a greater number of ADC pins, Bluetooth support, and more compared to the NodeMCU development board (its predecessor), though it comes at a slightly raised price tag of approximately $8 to $10.

ESP32-based development board

ESP32 is often considered a successor to the NodeMCU development board. It is yet another open source board designed by Espressif specifically for prototyping mobile devices, wearable electronics, and IoT applications.

The main upgrade over the last generation chip is a hybrid Wi-Fi and Bluetooth chip, which can help you provide an additional connection protocol to your application. Note that the ESP32 dev boards have a single antenna, so only one of the two protocols can be used at a given time.

The following figure shows what the development looks like. The form factor is similar to the NodeMCU development board but with some notable differences:

Figure 3.7 – ESP32 DEVKIT V1

We will be using this particular model of the ESP32 board in this book, but the information given here will be also applicable for its other versions as they are all powered by the same chip.

An overview of the ESP32 specifications is as follows:

- The ESP32 board is a dual-core 32-bit processor.
- It has Wi-Fi and Bluetooth built-in.
- Its clock frequency is up to 240 MHz and it has a 512 kB SRAM.

- This particular board has 30 GPIO pins, with 15 in each row.
- It also has a wide variety of peripherals available, such as capacitive touch, ADCs, DACs, UART, SPI, I2C, and much more.

In the next section, we will cover the technical specifications of the board in detail.

Technical specifications

The following table shows a detailed specification chart for this device. It covers all the important features and peripherals of the ESP32 development board (ESP32 DEVKIT V1 in particular):

Category	Specifications
MCU	ESP32 Xtensa 32-bit dual-core CPU
Wi-Fi	Yes, HT40
Bluetooth	Bluetooth 4.2 and below
Clock Frequency	160 MHz
SRAM	512 KB
Flash Memory	SPI Flash, up to 16 MB
GPIO Pins	30
PWM (Hardware/Software)	1 hardware, 16 channel software
ADC/CAN	12-bit ADC (18 channels)/ 1
SPI/I2C/I2S/UART	3/2/2/3
Onboard Sensors	Touch and Temperature
Working Temperature	-40 to 125 °C

Table 3.1 – Technical specifications of ESP32 DEVKIT V1

Next, we will cover the pinout of the ESP32 development board in detail, as we did for the NodeMCU development board. There are some additional features that this board boasts over its predecessor in terms of GPIO pins as well.

ESP32 GPIO pinout and pin configurations

The following diagram shows that there are numerous types of pins on the board:

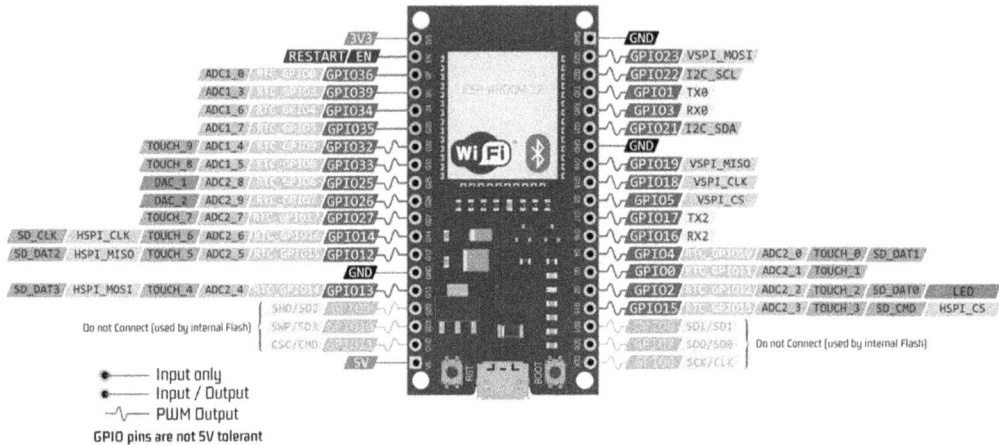

Figure 3.8 – ESP32 development board pinout diagram

A single pin is multiplexed so that it can perform multiple tasks. These can be decided on using its multiplexing select registers. For now, we will stick to the simple parts and provide an overview of the different pins there are:

- There are two **Power pins**: VIN and 3.3 V.

  The VIN pin is used to provide external power to the board, while the 3.3 V pin powers the sensors connected to the board.

- There are two **Ground pins** on the board that can be used for multiple purposes.

- There are 15 accessible **ADC pins**, which are marked as $ADCx\_y$ in the preceding diagram. These pins are used to collect analog data from various sensors or other sources.

- There are 25 **GPIO pins** (purple-colored and multiplexed) that the user can use to connect various digital data sensors to the board. However, there are always some essential pins that will be needed for other applications, so using all 25 pins at once is very rare, but possible.

- The ESP32 board even has two **Digital to Analog (DAC)** converter pins.

- There are three **UART channels** in total, out of which one is used by the micro USB port and the remaining two can be accessed through the U0 and U2 Tx and Rx pins.

- The board also has **Serial Peripheral Interface** (**SPI**) pins, which include the MOSI, MISO, SCK, and CS pins. Moreover, it also provides I2C protocol support by using the provided SDA and SCL pins.

- Nine pins can be used to access the on-chip **Touch sensor**. These are marked as Touch0 to Touch9 in the preceding pinout diagram.

- There are two **XTAL pins**, which can be used to connect an external crystal (to provide a clock signal) to the ULP processor.

- Finally, there are the **SensVP** and **SensVN** pins, which are dedicated to measuring small DC signals (for example, from a thermocouple).

This concludes the pinout diagram description for the ESP32 board. If you wish to learn more, please check out the official documentation of the ESP32 chip, which is provided on the official Espressif site:

```
https://docs.espressif.com/projects/esp-idf/en/latest/esp32/
```

Arduino IDE setup for the ESP32 development board

When it comes to programming, the ESP32 board can easily be programmed through the Arduino IDE, which is an open source development environment for different development boards (by default, it can be only used for Arduino boards).

Fortunately, we have to follow the same process we did for the NodeMCU board, but with a few minor changes. Some additional steps need to be followed to program the ESP32 board:

- **Change 1**: Follow *Steps 1* to *3* for the NodeMCU board as-is. In *Step 4*, please copy the `https://raw.githubusercontent.com/espressif/arduino-esp32/gh-pages/package_esp32_index.json` link instead of the one specified previously. Note that to keep both, just put a comma (,), after the ESP8266 link and type in the link mentioned previously.

  Now, **Boards manager** knows where to look for all the necessary files that will allow us to program the ESP32 boards using the Arduino IDE.

- **Change 2**: When you open **Boards Manager** in the next step, instead of searching for `esp8266`, just search for `esp32` and install the necessary files for them. It will take around 10 minutes to download and install all the necessary files.

- **Change 3**: Finally, instead of choosing the NodeMCU as the board, we must choose the **ESP32 Board** option.

  You would most probably have an ESP32 DEVKIT V1 as it is the most popular and cheapest option of all. Choose the **DOIT ESP32 DEVKIT V1** option from the **ESP32 Boards** section.

  Now, you should be able to successfully load the Blink program onto your ESP32.

> **Important Note**
>
> In some ESP32 boards, the program won't start uploading automatically once you press the **Upload** button in the IDE. It will keep trying to detect the board. There is a very simple fix for this issue.
>
> Once you have pressed the **Upload** button, just press and hold the **BOOT** button on the ESP32 kit. It will automatically detect the board and start uploading.

With that, we have introduced two very useful and powerful development boards developed specifically for **Internet of Things (IoT)** applications.

Please note that we will be using these boards throughout this book. In fact, in the later chapters, these boards will act as the sensor nodes for our projects, so you need to become familiar with these devices.

In the next section, we will be creating our first project, wherein we will use an external development board as an MQTT client that will establish a connection to your Raspberry Pi MQTT broker. Here, we will do the following:

- Set up a NodeMCU device as an MQTT client.
- Connect this client to our Raspberry Pi MQTT broker.
- Send dummy data from the NodeMCU device to our Pi MQTT broker and control the NodeMCU's onboard LED through MQTT.

Let's get started.

Mini-project 1: NodeMCU as an MQTT client

This is the first project we will be doing related to MQTT and our Raspberry Pi broker.

First, we will start by setting up the NodeMCU board. No external connections need to be made as we are only controlling the on-chip LED. This will be divided into two parts:

- Node MCU setup and code explanation
- Raspberry Pi setup and project demonstration

We will first set up our NodeMCU development board for this project.

Part 1 – NodeMCU development board setup

In this section, we will program our NodeMCU board to act as an MQTT client and control its onboard LED using an external device (your home computer, in this case).

For this, we will write a sketch that gives us access to the features we need. We will be using the **pubsub** library for this purpose. The code we will be using will do the following:

1. First, it will connect to an MQTT server.

2. Once the connection has been established, it will publish "hello world" to the **outTopic** topic every 2 seconds.

3. Once the connection has been established, it will subscribe to the **inTopic/LED** topic, printing out any messages it receives.

4. It will receive messages from the subscribed topics and assume that the received payloads are strings, not binary. If the first character of **inTopic** is a 1, it will be programmed to switch the onboard ESP LED on; otherwise, it will switch it off.

5. It will reconnect to the server if the connection is lost using a blocking reconnect function.

Code explanation

Now, let's go through the code in parts so that we can understand it better:

1. The first two lines of code import the required libraries. The `ESP8266WiFi` library is used to access Wi-Fi networks, which grants the board internet access. The `PubSubClient` library is the MQTT client library, which helps us run an MQTT client on the board:

   ```
   #include <ESP8266WiFi.h>
   #include <PubSubClient.h>
   ```

2. The next three lines are constant variable initializations for the *Wi-Fi name, password, and the MQTT (our Raspberry Pi's) IP address*. The Raspberry Pi's IP address is the IP address of the broker.

   The following few lines deal with various object and variable initializations that will be required later in our code:

   ```
   const char* ssid = "wifi_name";
   const char* password = "wifi_password";
   const char* mqtt_server = "ip_of_raspberry_pi";

   WiFiClient espClient;
   PubSubClient client(espClient);
   unsigned long lastMsg = 0;
   #define MSG_BUFFER_SIZE     (50)
   char msg[MSG_BUFFER_SIZE];
   int value = 0;
   ```

3. `setup_wifi` is a custom function that's used to connect to the Wi-Fi network. We provided the network's credentials as constant variables:

```
void setup_wifi()
{
  delay(10);
  // We start by connecting to a WiFi network
  Serial.println();
  Serial.print("Connecting to ");
  Serial.println(ssid);
  WiFi.mode(WIFI_STA);
  WiFi.begin(ssid, password);
  while (WiFi.status() != WL_CONNECTED)
  {
    delay(500);
    Serial.print(".");
  }

  randomSeed(micros());

  Serial.println("");
  Serial.println("WiFi connected");
  Serial.println("IP address: ");
  Serial.println(WiFi.localIP());
}
```

4. The `callback` function is used to print out the data that's received on the subscribed MQTT channels. In this case, we are subscribed to a topic called *inTopic*. We will be sending data to this to control the onboard LED. We have programmed the function so that if it receives 1 on the topic, the LED turns on, while if it receives 0 on the topic, it turns off:

```
void callback(char* topic, byte* payload, unsigned int
length)
{
  Serial.print("Message arrived [");
  Serial.print(topic);
  Serial.print("] ");
```

```
  for (int i = 0; i < length; i++) {
    Serial.print((char)payload[i]);
  }
  Serial.println();
// Switch on the LED if an 1 was received as first
character
  if ((char)payload[0] == '1')
  {
    digitalWrite(BUILTIN_LED, LOW);
  }
  else
  {
    digitalWrite(BUILTIN_LED, HIGH);
  }
}
```

- The `reconnect` function is used to reconnect to the MQTT broker in case there is a problem and the board disconnects. We must create a random client ID so that there is no possibility of a duplicate client ID. We use the `random` function and create a 16-byte random hex value and append it with the `ESP8266Client-` string. The reasons for disconnection may include internet connection failure, duplicate client IDs, and so on.

- After that, we will be using the pub-sub client library's connect, publish, and subscribe functions (the main MQTT functions) for specific tasks. Let's take a look at these now.

- **Connect function**

  First, we will attempt to establish a connection with the MQTT broker with the generated client ID using the `connect` function. Please note that you can have other arguments for this function as well (for instance, if you have security enabled, you will need to send the user ID and password as well).

  All these parameters were discussed in detail in *Chapter 2, MQTT in Detail*, when we discussed the theory of MQTT in detail. Information about this function is provided in the following screenshot:

boolean **connect** (clientID, [username, password], [willTopic, willQoS, willRetain, willMessage], [cleanSession])

Connects the client.

Parameters
- clientID const char[] - the client ID to use when connecting to the server
- Credentials - *(optional)*
  - username const char[] - the username to use. If NULL, no username or password is used
  - password const char[] - the password to use. If NULL, no password is used
- Will - *(optional)*
  - willTopic const char[] - the topic to be used by the will message
  - willQoS int: 0,1 or 2 - the quality of service to be used by the will message
  - willRetain boolean - whether the will should be published with the retain flag
  - willMessage const char[] - the payload of the will message
- cleanSession boolean *(optional)* - whether to connect clean-session or not

Returns
- false - connection failed
- true - connection succeeded

Figure 3.9 – Pub-sub client connect function explained

- **Publish function**

  Once the connection to the broker has been established, we publish a *hello world!* message on the *outTopic* topic to acknowledge that the connection has been established. The publish function performs this task. Information about this function is provided in the following screenshot:

boolean **publish** (topic, payload, [length], [retained])

Publishes a message to the specified topic.

Parameters
- topic const char[] - the topic to publish to
- payload const char[], byte[] - the message to publish
- length unsigned int *(optional)* - the length of the payload. Required if payload is a byte[]
- retained boolean *(optional)* - whether the message should be retained
  - false - not retained
  - true - retained

Returns
- false - publish failed, either connection lost or message too large
- true - publish succeeded

Figure 3.10 – Pub-sub client publish function explained

- **Subscribe function**

Finally, once we have the connection, we will `subscribe` to the `inTopic/LED` topic to control the onboard LED on the NodeMCU board. The function just takes the topic name as an argument for this project. However, we can also set the QoS (which was discussed in *Chapter 2, MQTT in Detail*, as well) value for this topic. Information about this function is provided in the following screenshot:

boolean **subscribe** (topic, [qos])

Subscribes to messages published to the specified topic.

Parameters
- topic const char[] - the topic to subscribe to
- qos int: 0 or 1 only *(optional)* - the qos to subscribe at

Returns
- false - sending the subscribe failed, either connection lost or message too large
- true - sending the subscribe succeeded

Figure 3.11 – Pub-sub client subscribe function explained

If you still need to dive deeper into the pub sub-client library, you can find the complete documentation for this at `https://pubsubclient.knolleary.net/api#connect`.

Note that this function tries to connect to the MQTT broker every 5 seconds:

```
void reconnect()
{
  // Loop until we're reconnected
  while (!client.connected()) {
    Serial.print("Attempting MQTT connection...");
    // Create a random client ID
    String clientId = "ESP8266Client-";
    clientId += String(random(0xffff), HEX);
    // Attempt to connect
    if (client.connect(clientId.c_str())) {
      Serial.println("connected");
      // Once connected, publish an announcement...
      client.publish("outTopic", "hello world");
      // ... and resubscribe
      client.subscribe("inTopic/LED");
```

```
      } else {
        Serial.print("failed, rc=");
        Serial.print(client.state());
        Serial.println(" try again in 5 seconds");
        // Wait 5 seconds before retrying
        delay(5000);
      }
    }
  }
```

5. This is the Arduino `setup` function. Here, we set the built-in LED pin mode to output. Then, we initiate a serial connection with a baud rate of `115200`. Next, we call the `setup_wifi` function to connect to the Wi-Fi network.

    Next, we connect to the MQTT server, which in our case is the Raspberry Pi. Then, we specify the `callback` function using the `setCallback` function of the PubSubClient library. Finally, we subscribe to the `inTopic/LED` topic so that we can control the onboard LED:

```
    void setup()
    {
      pinMode(BUILTIN_LED, OUTPUT);
      // Initialize the BUILTIN_LED pin as an output
      Serial.begin(115200);
      setup_wifi();
      client.setServer(mqtt_server, 1883);
      client.setCallback(callback);
      client.subscribe("inTopic/LED");
    }
```

The final `loop` function runs indefinitely. First, we check if the MQTT client is disconnected and if so, we run the reconnect function. After that, we publish a *Hello world x* message every 2 seconds to the `outTopic` topic, where *x* is a variable whose value keeps incrementing.

In addition to this, we already have the callback function defined. So, if we send a 0 or 1 to `inTopic`, we can control the LED on the NodeMCU board:

```
    void loop() {
      if (!client.connected()) {
        reconnect();
      }
```

```
    client.loop();
    unsigned long now = millis();
    if (now - lastMsg > 2000) {
      lastMsg = now;
      ++value;
      snprintf (msg, MSG_BUFFER_SIZE, "hello world #%ld",
  value);
      Serial.print("Publish message: ");
      Serial.println(msg);
      client.publish("outTopic", msg);
    }
  }
}
```

With that, we have explained the code. You can find the complete code for this project in this book's GitHub repository, along with comments to guide you through. Just upload this code to the NodeMCU board to complete the NodeMCU setup.

Now, let's set up the Raspberry Pi.

Part 2 – Raspberry Pi setup

This is the easy part of this tutorial. All we need to do is to test and demonstrate that we can control the onboard LED of the NodeMCU board wirelessly through MQTT.

We will use the Raspberry Pi as an MQTT client to control the NodeMCU's onboard LED. Follow these steps:

1. Open a Terminal and start the MQTT broker on the Pi if it is not already active. It is highly unlikely that it would be inactive if you have followed the steps provided in the previous chapters.

2. Now, open two new Terminal windows. We will use one of them as an MQTT subscriber that subscribes to outTopic and the other as an MQTT publisher, which will let us control the onboard LED of the NodeMCU board.

3. In one Terminal window, type the following command:

    ```
    mosquitto_sub -v -t outTopic
    ```

    This will subscribe to outTopic, which the NodeMCU board has been sending the dummy messages to. They should start appearing on the Terminal every 2 seconds.

4. In the second Terminal, type the following two commands one after the other:

    ```
    mosquitto_pub -t inTopic/LED -m 1
    mosquitto_pub -t inTopic/LED -m 0
    ```

The first command will turn on the NodeMCU's onboard LED, while the second command will turn it off.

That's all we need to do to set up the Raspberry Pi. The following screenshot shows what the final screen of the Raspberry Pi should look like. You can also see the Arduino IDE Serial Monitor output in the image, which logs all the published and received messages:

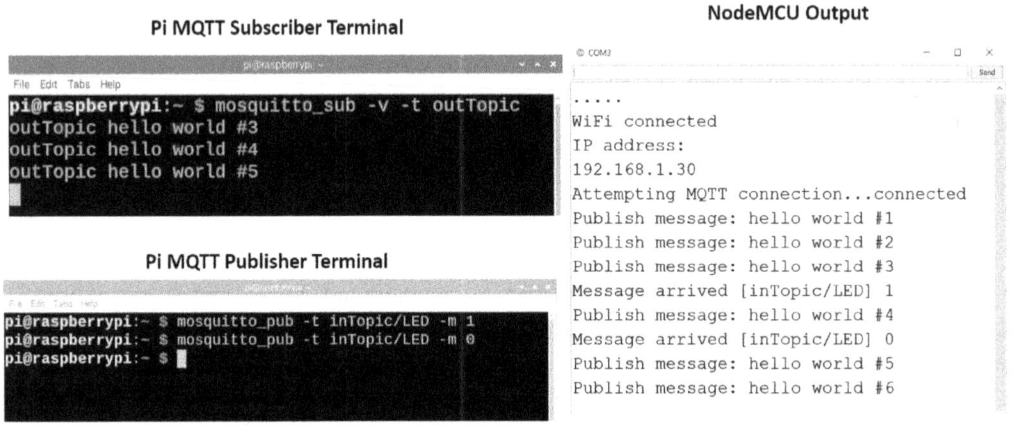

Figure 3.12 – Mini-project output demonstration – Raspberry Pi

Great! You've just completed your very first mini-project and controlled a peripheral of an MQTT client wirelessly using the MQTT protocol. This also marks the end of this chapter. Now, let's summarize what we've learned.

Summary

This chapter was solely dedicated to the ESP development boards that we will be using throughout this book, even in the full-scale projects that we will be creating in the upcoming chapters. First, we covered the *NodeMCU development board*, including its technical specifications, and its GPIO configuration, before learning how to set up the Arduino IDE for this board and flashing our very first program. Next, we covered the same topics for the *ESP32 development board*, which is often considered a successor of the latter with some additional features. Finally, we *built our very first mini-project*. We turned our NodeMCU board into an MQTT client and controlled its onboard LED wirelessly using the MQTT communication protocol.

In the next chapter, we will look at another important component that will be very useful later in this book when we cover two full projects: **Node-RED**.

We will use this to create *dashboards* for our projects and learn how we can store the received MQTT data in a SQL database later in this book.

4

Node-RED on Raspberry Pi

This chapter will get you acquainted with very popular software for the **Raspberry Pi** – **Node-RED**. It is browser-based low-code programming software that allows beginners to create APIs and control the Pi hardware by creating flows, a connected component created by wiring several nodes together to perform a specific task. This chapter has four main sections:

- Introduction to Node-RED
- Node-RED first-time installation, setup, and demonstration
- Node-RED MQTT components and dashboard setup
- Mini project 2 – controlling a NodeMCU LED from the Node-RED dashboard

So, let's start with a basic introduction to what exactly Node-RED is.

Introduction to Node-RED

The website of Node-RED (`https://nodered.org/`) gives a perfect introduction to the purpose of using the software:

Node-RED is a programming tool for wiring together hardware devices, APIs and online services in new and interesting ways.

It provides a browser-based editor that makes it easy to wire together flows using the wide range of nodes in the palette that can be deployed to its runtime in a single click.

Hence, we can say that Node-RED is a UI-based programming tool that can be used to create various applications, which include hardware device control (Raspberry Pi GPIO access is also available), flow-based API development, and so on. We can run multiple flows at once and each of them will run independently.

Moreover, Node-RED allows the easy setup of several additional services, which would otherwise require a lot of work and experience. One such example is setting up a SQL database to store all the data arriving in Node-RED through any communication protocol, be it **hardware-based** (**GPIO**) or **software-based** (**MQTT**).

This software is so useful, especially for a device such as the Raspberry Pi, that it comes pre-installed when you flash the Raspberry Pi OS version with the recommended software. Additional Node-RED developments were made specifically for the Raspberry Pi. These include the following:

- You can access and control the Raspberry Pi GPIO from Node-RED.

- Due to the limited memory of the Raspberry Pi, you will need to start Node-RED with an additional argument to tell the underlying Node.js process to free up unused memory sooner than it would otherwise.

To do this, you should use the alternative `node-red-pi` command and pass in the `max-old-space-size` argument. Please note that this is browser-based software so when you run this locally on the Pi, you can access the console window from any device connected to the same network as the Raspberry Pi.

This section has given us an informative introduction to Node-RED. In the next section, we will first install and set up Node-RED on our Raspberry Pi and then create two simple flows to get some basic knowledge on how to use this software.

Node-RED first-time installation, setup, and demonstration

This section will cover in detail how to install and set up Node-RED on the Raspberry Pi. After that, two simple demonstration applications will be covered, which will show us the power of Node-RED and how it can be utilized to its full extent.

Node-RED installation

Installing and setting up Node-RED on your Raspberry Pi is a straightforward process. Please note that this section assumes that you have already set up your Raspberry Pi with the latest version of the Raspberry Pi OS with all the initial configurations. If that is not the case, please refer to *Chapter 1, Introduction to Raspberry Pi and MQTT*, for that.

First, access your Raspberry Pi as we will require the terminal window at the very least to install Node-RED on it. There are two ways to access your Pi:

- **Through the desktop interface**: It requires a monitor, keyboard, and mouse connected to the Pi.

- **Through Secure Shell Protocol (SSH)**: SSH has to be enabled from Pi configuration to use this option. You can even use **VNC** (which stands for **Virtual Network Computing**) to access the whole desktop interface through your PC and use your PC's mouse and keyboard.

You will need the terminal window regardless of the method you use for this setup process. Please note that you can use any software to SSH into your Raspberry Pi. *Figure 4.1* shows the terminal window after you SSH into your Pi using Putty:

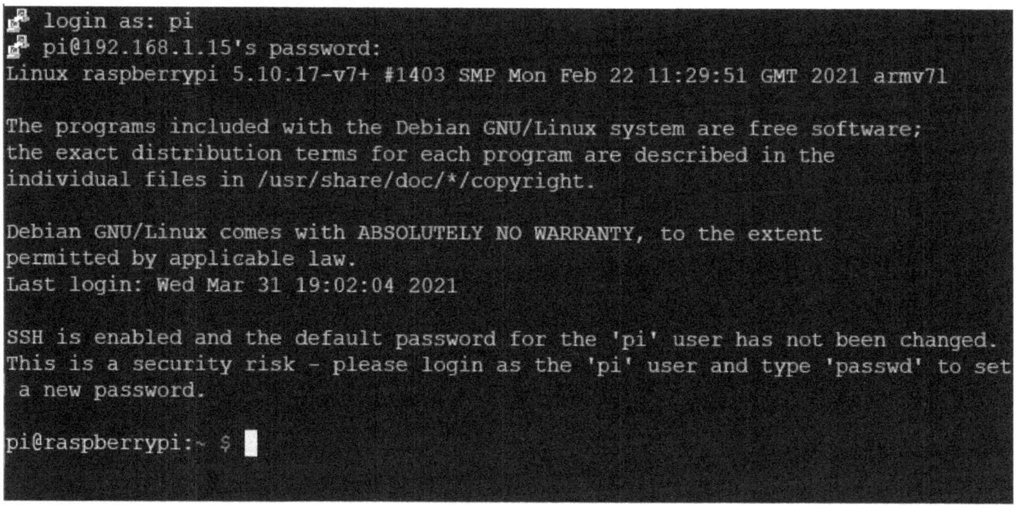

```
login as: pi
pi@192.168.1.15's password:
Linux raspberrypi 5.10.17-v7+ #1403 SMP Mon Feb 22 11:29:51 GMT 2021 armv7l

The programs included with the Debian GNU/Linux system are free software;
the exact distribution terms for each program are described in the
individual files in /usr/share/doc/*/copyright.

Debian GNU/Linux comes with ABSOLUTELY NO WARRANTY, to the extent
permitted by applicable law.
Last login: Wed Mar 31 19:02:04 2021

SSH is enabled and the default password for the 'pi' user has not been changed.
This is a security risk - please login as the 'pi' user and type 'passwd' to set
 a new password.

pi@raspberrypi:~ $
```

Figure 4.1 – Raspberry Pi terminal access through SSH (via Putty)

Next, we need to make sure that the Raspberry Pi OS is up to date. For that, just run the two commands given next in the order they are given. This will take some time, so I would suggest you go out for a walk or grab a snack in the meantime:

```
sudo apt update
sudo apt upgrade
```

You may need to restart your Pi for the updates to be installed. After this is complete, we can proceed to the actual software installation. As mentioned before, Node-RED comes pre-installed with the full OS image. But this may be using an older version of Node.js. To install/upgrade to the latest version of Node-RED (along with the dependencies), please use the following command:

```
bash <(curl -sL https://raw.githubusercontent.com/node-red/
linux-installers/master/deb/update-nodejs-and-nodered)
```

This will walk you through the whole installation process, which will also install all the required dependencies for Node-RED. Just press Y to start the installation process (*Figure 4.2*):

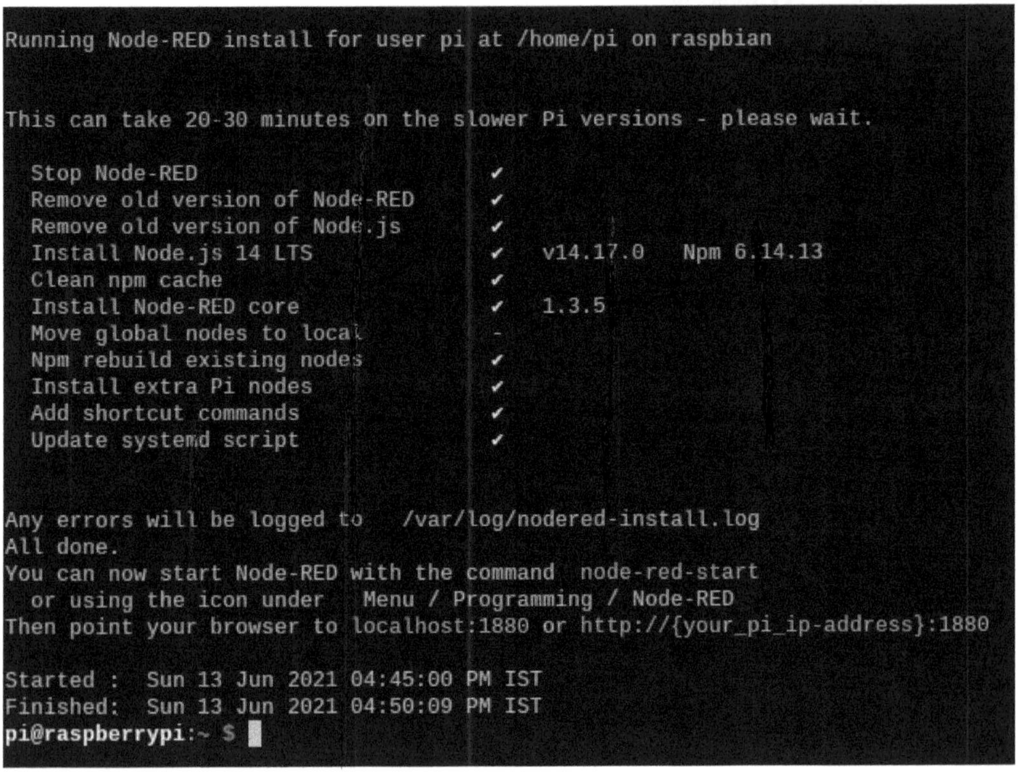

```
Running Node-RED install for user pi at /home/pi on raspbian


This can take 20-30 minutes on the slower Pi versions - please wait.

  Stop Node-RED                        ✔
  Remove old version of Node-RED       ✔
  Remove old version of Node.js        ✔
  Install Node.js 14 LTS               ✔    v14.17.0    Npm 6.14.13
  Clean npm cache                      ✔
  Install Node-RED core                ✔    1.3.5
  Move global nodes to local           -
  Npm rebuild existing nodes           ✔
  Install extra Pi nodes               ✔
  Add shortcut commands                ✔
  Update systemd script                ✔


Any errors will be logged to    /var/log/nodered-install.log
All done.
You can now start Node-RED with the command  node-red-start
  or using the icon under   Menu / Programming / Node-RED
Then point your browser to localhost:1880 or http://{your_pi_ip-address}:1880

Started :  Sun 13 Jun 2021 04:45:00 PM IST
Finished:  Sun 13 Jun 2021 04:50:09 PM IST
pi@raspberrypi:~ $ █
```

Figure 4.2 – Node-RED installation logs

Next, we will see how we start Node-RED on our Raspberry Pi, now that we have successfully installed it.

Running Node-RED on your Pi for the first time

There are two ways to start Node-RED, one through the desktop interface and the other through a terminal window. To start it through the desktop interface, just open the Node-RED application through the following path: **Start Menu | Programming | Node-RED**. This will open a terminal window with the node-red-start command pre-configured to run.

To start it through the terminal, just open a new terminal window and type in the following command:

```
node-red-start
```

Once you start Node-RED through any of the mentioned methods, wait a few seconds for the initialization process to complete. Once that is done, you will see a message flash on the terminal window saying: **Node-RED has started, point a browser at <—IP address of Node-RED—>**. This means that Node-RED is now active on the given IP address. The Node-RED server runs on the same IP address as that of your Raspberry Pi and on port 1880.

You can access Node-RED through any device browser that is connected to the same Wi-Fi network as your Raspberry Pi. This makes this service easily accessible across multiple platforms with the host being the Pi. To open Node-RED on any device, all you need to do is the following:

1. First, connect to the Wi-Fi network to which your Raspberry Pi is connected (your device could be a PC, laptop, mobile phone, or even a smart TV!).

2. Then, open a browser window.

3. In the URL bar, type in your Raspberry Pi's IP address (which you use to access it via VNC), followed by a colon, and then the port number for Node-RED, which is 1880. For example, if your Pi's IP address is 192.168.0.23, then the URL you need to type in will be 192.168.0.23:1880.

In the next section, we will go through a crash course on all the basic functionalities and features Node-RED has to offer. The section will further cover two very simple and basic applications that will utilize the various functionalities of Node-RED.

Node-RED crash course

Now that we have Node-RED installed on our Raspberry Pi, we can further explore all the features and functionalities it has to offer. We will go about that in the following order:

*   What do we see on the home screen?

*   Understanding the following terms–node and flow

*   Additional features

*   Creating the *Hello World* flow

*   Controlling an LED connected to the Raspberry Pi through Node-RED

Please note that this will give you an overview of how Node-RED works on the Raspberry Pi. So, let's get started with the first point.

What do we see on the home screen?

When you open Node-RED on any web browser, you will see the home screen shown in *Figure 4.3*.

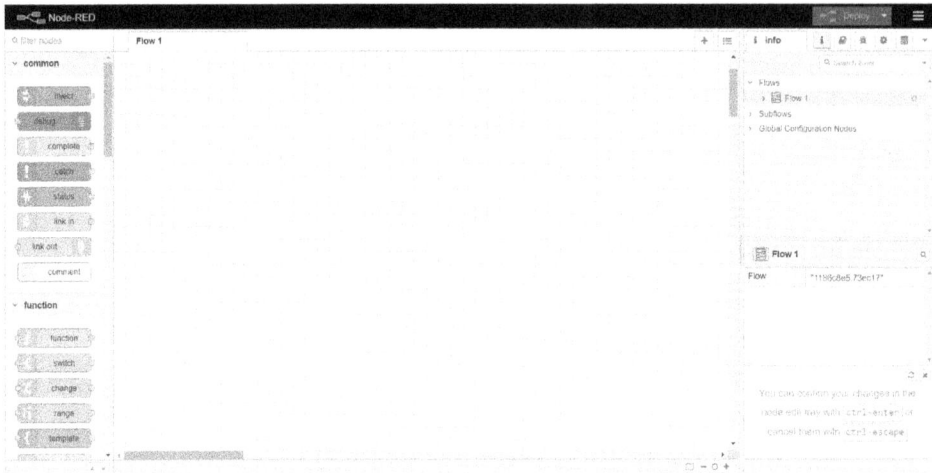

Figure 4.3 – Node-RED home screen

Now, to understand this screen better, we will divide the main screen into three components and go through each separately (*Figure 4.4*). The names of these components are shown next:

1. Node palette
2. Workspace
3. Control panel

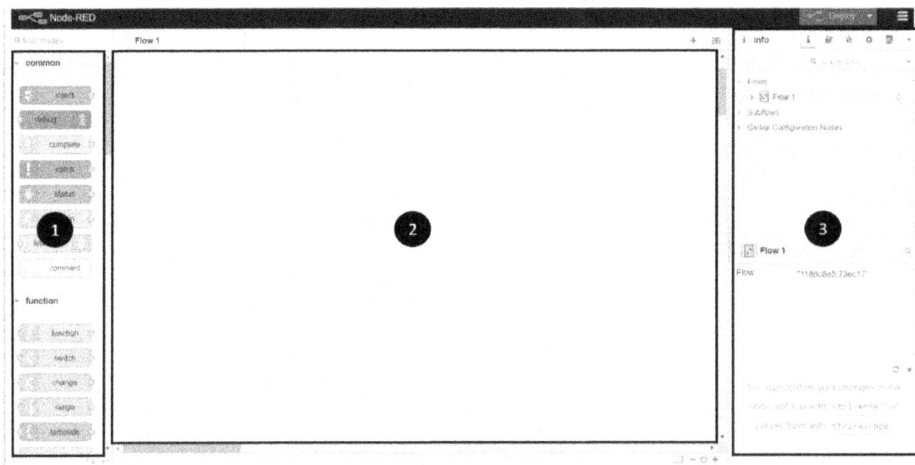

Figure 4.4 – Three components of the Node-RED home screen

Each part has its purpose and that is what we will be discussing in this section.

1 – Node palette

The node palette is a very important, essential component. It has all the nodes that can be used to create different flows with application-specific functionalities in Node-RED. We will discuss these terms in detail in the next subsection.

The palette is further segregated into different sections based on the category a particular node belongs to. This helps the user search for the right node easily. Each node has its required configuration. Some nodes can be used directly just by dropping them into the workspace and some require additional setup.

2 – Workspace

A major part of the home screen is occupied by the workspace. This is the area where we create our flows. We drag all the nodes we require for a particular application from the palette to the workspace and connect them in a particular order to create a flow (which can be interpreted as a program).

3 – Control panel

This is the last component of the Node-RED home screen. This panel has several tabs, each serving a particular purpose. We will be discussing the important tabs, as follows:

- It has the **Info** tab, which contains information such as how many flows the workspace has, what nodes are used in each flow, and so on.

- The **Help** tab gives information about each node, which includes several things, such as a short description of the node, how to use the node in a flow, and so on.

- The **Debug** tab opens the **Debugging** panel where the debug information of each node is visible (it can be configured in a flow using the debug node).

- The **Dashboard** tab is also accessible through this component. It gives information about the components and the dashboard can be opened from this section (note that this tab is not visible in the preceding figure as we need to install the Node-RED dashboard extension).

Understanding the following terms–node and flow

The following are explanations of the terms:

- **Node**: We will use a very simple analogy to understand what a node is. Whenever you write code for an application, you create several functions to make the code more readable and efficient. The nodes are different functions, with each node having particular functionality.

- **Flow**: Continuing with the preceding analogy, when we use multiple nodes and join them to create a group of nodes performing a specific task, this group is termed a *flow*. This is like the code written for that task but instead of code, we use Node-RED's drag and drop feature to create a flow performing the same task.

Additional features

In addition to the basics that we just covered, there are ample additional features. A major feature is the ability to "install external node extensions."

That opens a lot of new development opportunities, which include connecting Node-RED to databases, external services such as **IFTTT** (**If This Then That**), Twilio, and so on. The main advantage is that in most cases, no external tech stack knowledge is required. The setup is minimal, such as entering some required information. For example, to connect to a SQL database through the MySQL extension (this will be covered in detail in later chapters), you have to just enter the IP address on which the database is hosted, its port, database name, and user credentials. There is no need to write a single line of code.

There is an extensions library hosted and managed by www.nodered.org. We can access that through the Node-RED home page through the additional options, which can be accessed by clicking the three-line icon on the top-right portion of the screen (*Figure 4.5*):

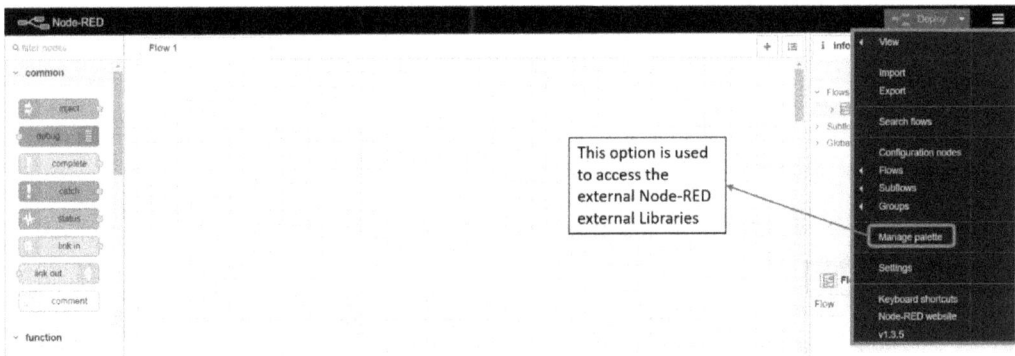

Figure 4.5 – Additional options section on the Node-RED screen

This concludes this section. Next, we will be creating two simple flows, which will help us understand how to operate Node-RED.

Creating the "Hello World" flow

In this section, we will be creating our very first flow with Node-RED. This flow will perform a very simple task:

Every time we trigger the flow, the current timestamp value will be printed on the debug screen (in the control panel).

To create this flow, we will require two nodes: the **inject** node and the **debug** node. Just drag these nodes from the node palette onto the workspace once. When this is done, your screen will look like what's shown in *Figure 4.6*.

Important Note

We need to deploy every time we make any changes in any of the Node-RED flows. Every added or updated node will have a light-blue dot on top, indicating that these changes will only be reflected when we redeploy the flows (dots visible in *Figure 4.6*).

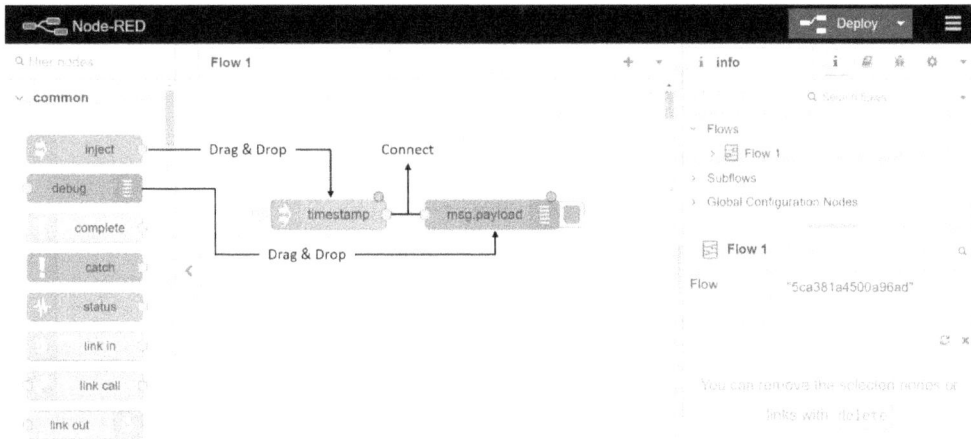

Figure 4.6 – Add the nodes required to create our first flow

The next step is to connect these nodes. To do that, just take your cursor to the gray dot on the **inject** node and then drag the line that emerges when you click and hold the left mouse button to the **debug** node's gray dot. This will connect these nodes.

This is all for this flow as no additional setup is required for either node. Now, if you click the **inject** node, nothing will happen. To use this flow, we need to deploy the updated flows. To deploy the flows, all we need to do is click the **Deploy** button situated on the top-right side of the screen. Once that is done, we are all set to test our first Node-RED flow!

Testing this flow is very easy. Just open the **Debug** tab from the control panel and then click the **inject** node. Clicking the blue button on the **inject** node will trigger the flow and the timestamp value will be printed on the **Debug** tab (*Figure 4.7*):

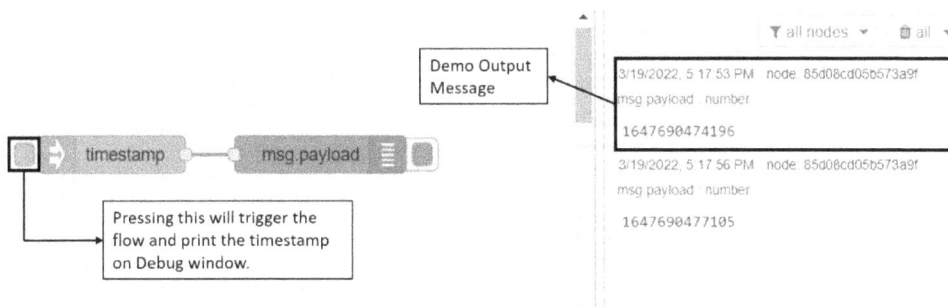

Figure 4.7 – First flow's output demonstration

This was the basic version of this flow. There are a few improvements that can be applied to this flow. For instance, instead of having to manually click the **inject** node to trigger the flow every time, you can automate this process so that the flow is automatically triggered after a certain interval. Moreover, instead of printing the timestamp on the **Debug** tab, we can print anything we want. So, we will modify this flow so that it automatically prints the infamous **Hello World!** string after every 10 seconds. All we need to do is make some changes in the **inject** node.

To do that, just double-click on the **inject** node, which will open an edit window. There are several customizations possible that differ from node to node. For our case, we will be making two changes:

1.   First, we will change the return value from **timestamp** to a custom string of our choice.

2.   Second, we will set up an interval of 10 seconds to trigger this node.

Refer to *Figure 4.8* to make these changes:

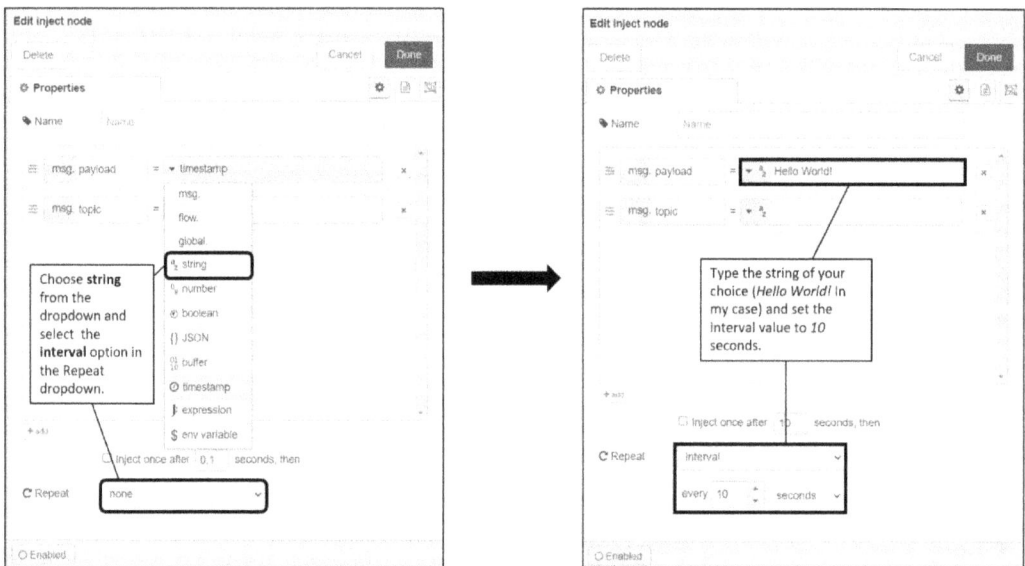

Figure 4.8 – Changes made in the inject node

Once these changes have been made, just redeploy the whole flow and see the new flow in action. Now, you will see the **Hello World!** string printed in the **Debug** tab every 10 seconds (*Figure 4.9*):

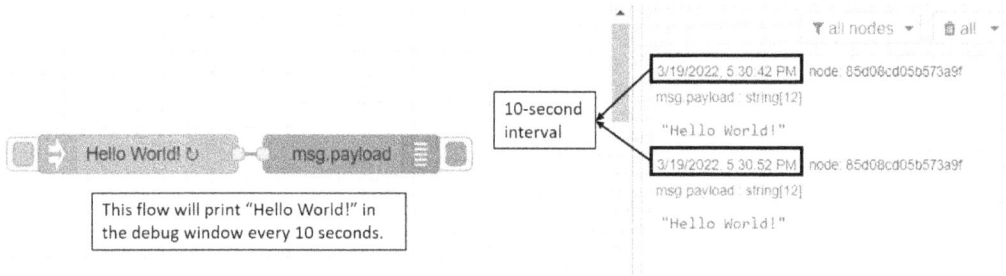

Figure 4.9 – Output of the updated flow

In the next section, we will create a flow to control the Raspberry Pi GPIO.

Controlling an LED connected to the Raspberry Pi through Node-RED

In this section, we will create a flow that will let us control a Raspberry Pi GPIO pin using the flow's UI. First, drag the following nodes onto the workspace window:

- Two **inject** nodes
- An **rpi – gpio out** node (found in the Pi-specific nodes section):

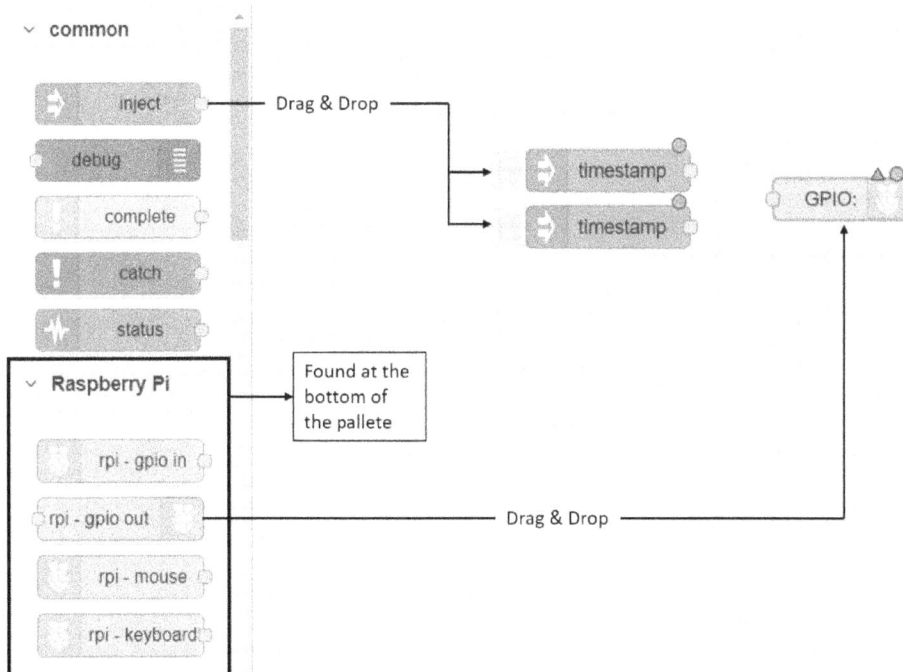

Figure 4.10 – Node-RED components for this flow

Next, we have to configure the nodes according to our requirements. These configurations include the following:

- Renaming the **inject** nodes and sending a Boolean `True` or `False` value according to the name given to it.

- Specifying which RPi GPIO we would like to configure with the GPIO node. In this case, we will choose GPIO18 (Pin 12).

Please refer to *Figure 4.11* for the necessary configurations:

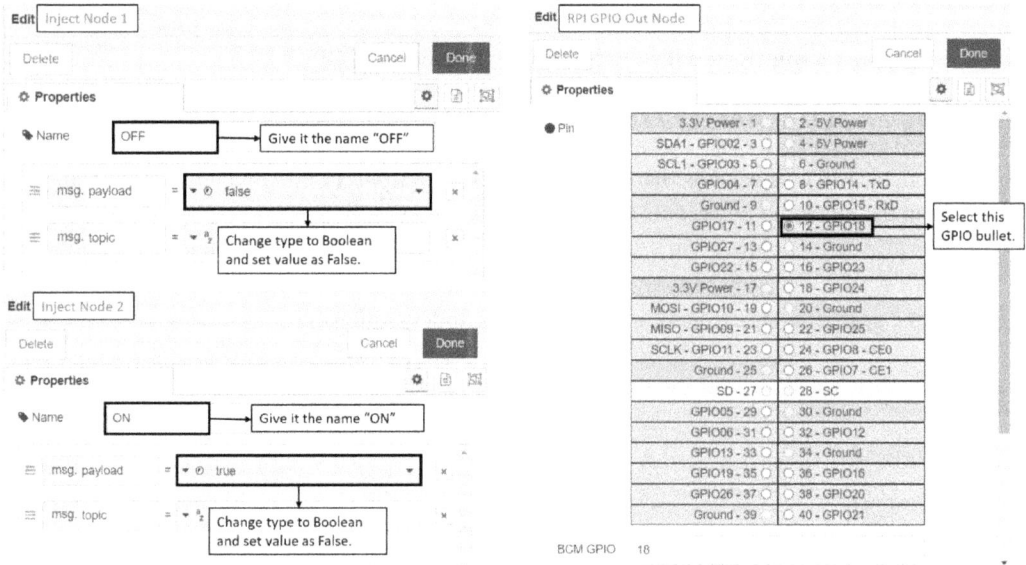

Figure 4.11 – Node configurations for each component of the flow

These were the required software configurations. In terms of hardware setup, you will need the following components:

- A 100–330-ohm resistor

- An LED

- A breadboard

Now, connect the hardware circuit using *Figure 4.12* as a reference.

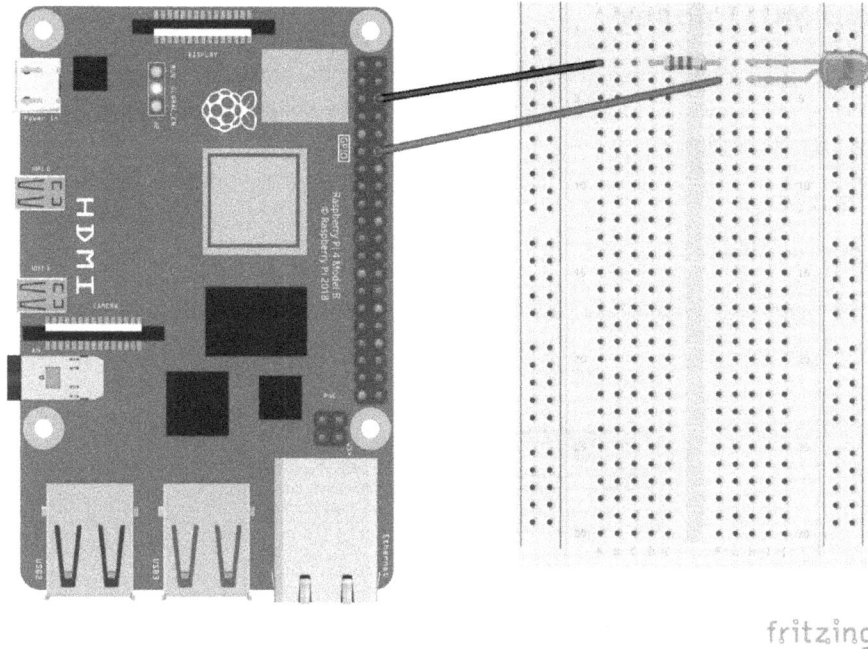

Figure 4.12 – Raspberry Pi LED connection schematic diagram

This completes all the required setup procedures. Now, just deploy the flow from the Node-RED home screen and you will now have the functionality to control the LED via Node-RED.

This is just a basic demonstration of Node-RED. Using different extensions and customizations, creating even complex projects is straightforward compared with the conventional coding alternatives that require programming knowledge. But it is important to note that for complex applications, you may need to write some custom functions in Node-RED that need to be coded in the JavaScript programming language (as Node-RED is based on Node.js). But for simple to intermediate-level projects, creating flows and customizing nodes is enough.

The next section will introduce you to the Node-RED dashboard. It is an interactive dashboard creation tool that will let you use different widgets such as switches, sliders, and gauges to create an interactive UI for the control and monitoring of your projects.

Node-RED MQTT components and dashboard setup

This section will start with an introduction to the dashboard and MQTT functionalities of Node-RED. That will be followed by a simple project in which we will connect an LED to a NodeMCU development board and demonstrate control capabilities through the Node-RED dashboard. So, let's get started.

Node-RED MQTT nodes

The MQTT communication protocol was discussed in detail in the previous chapters. In this section, we will be seeing how can we use MQTT nodes in Node-RED. In Node-RED, you can find two MQTT nodes in the network section:

- The **mqtt in** node
- The **mqtt out** node

Please refer to *Figure 4.13* to see where you would find the MQTT nodes in the node palette:

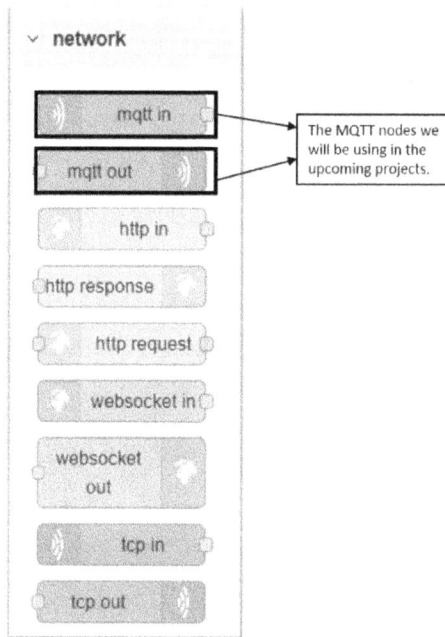

Figure 4.13 – MQTT nodes under the network section in the node palette

As the names suggest, the **mqtt in** node is to capture data coming from various topics (basically for monitoring) and the **mqtt out** node is to publish data on a particular topic (for control).

The following parameters are required and need to be configured for both nodes:

- **Server**: The IP of the MQTT server we need to use. In our case, we will be using the local MQTT server (localhost) hosted on our Raspberry Pi.
- **Topic**: The topic to which the node has to subscribe (to fetch data) or publish.
- **QoS**: This was discussed in detail in the last chapter.

Please take a look at *Figure 4.14*, which shows the edit options for both the in and out nodes:

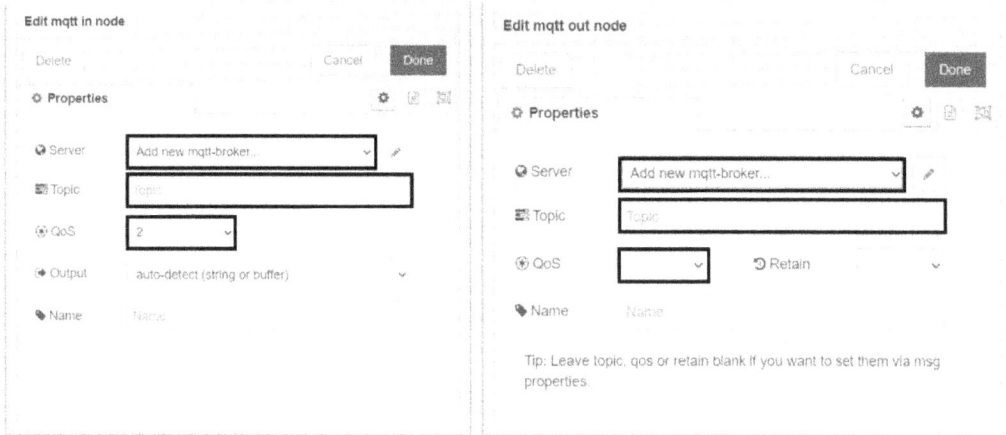

Figure 4.14 – MQTT in and out nodes' required configurations (marked)

There are several customizations when we want to add a new MQTT server. The main requirement is the server's IP address and the rest is optional (note that by default, the port is 1883, but this is different if we use SSL). The customizations include the MQTT version, client ID (autogenerated if kept blank), username and password in the **Security** tab, and some messages that are triggered when a client connects or disconnects from a topic.

Please refer to *Figure 4.15* to see the configurations available when adding a new MQTT broker for Node-RED MQTT nodes:

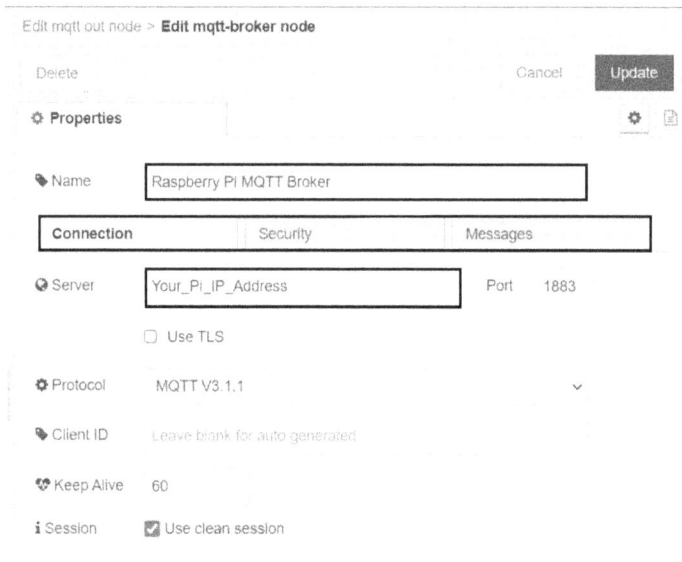

Figure 4.15 – Customizations available when adding a new MQTT broker

Next, we will cover the Node-RED dashboard installation, setup, and explanation. Please note that this is why we are using Node-RED, and hence, this is one of the most important sections of this chapter.

Node-RED dashboard

The Node-RED Dashboard is a module that provides a set of nodes in Node-RED to quickly create a live data dashboard. To learn more about the Node-RED Dashboard, you can check the following links:

Node-RED Dashboard site: `http://flows.nodered.org/node/node-red-dashboard`

GitHub: `https://github.com/node-red/node-red-dashboard`

There are two ways in which you can install this module. The first and the most straightforward way is to install it through the Node-RED UI. To do that, just go to options (the three horizontal bars on the top-right portion of the screen), then select **Manage palette,** and search `node-red-dashboard` in the **Install** section. You will see several options in the search results. Just click **install** for the first module (*Figure 4.16*). You have successfully installed the Node-RED Dashboard on your Pi:

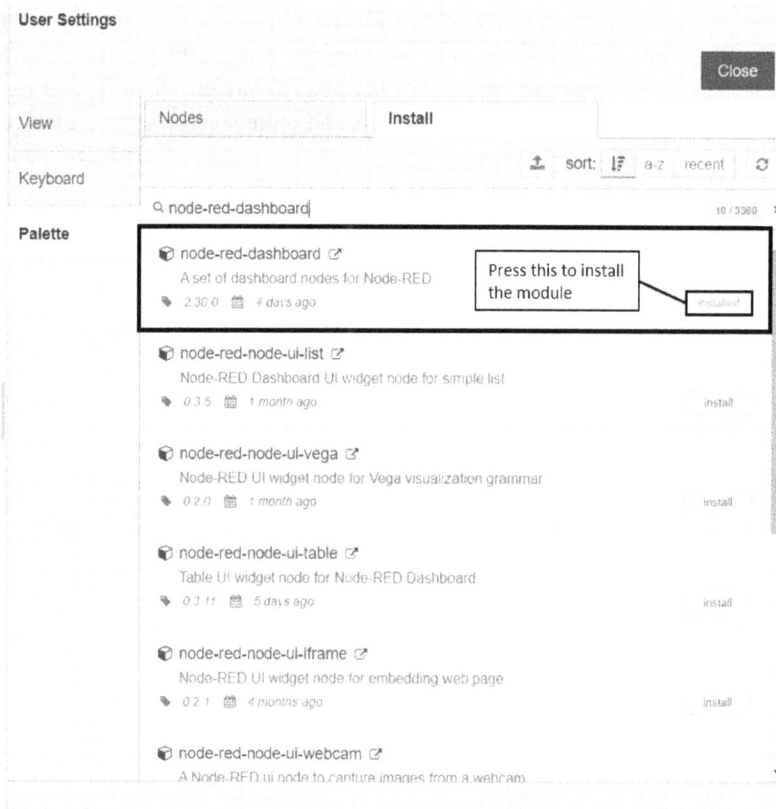

Figure 4.16 – node-red dashboard installation from the Node-RED UI

Alternatively, you can install `node-red-dashboard` through the terminal using **Node package manager (npm)**. For that, just type in the following commands in the same order:

```
node-red-stop
cd ~/.node-red
npm install node-red-dashboard
```

The first command will stop Node-RED if it is already running. After that, we need to move to the `node-red` directory to install any additional modules through npm, which is exactly what the second command does. After that, just install `node-red-dashboard` using the third command.

This will create a `node-red-dashboard` folder in the `node-red` directory where all the files related to the dashboard will be stored. To open the Node-RED UI, type your Raspberry Pi IP address in a web browser followed by `1880/ui`, as shown next:

```
http://<YourPi'sIPAddress>:1880/ui.
```

There will be two new additions on the Node-RED home screen. First, new nodes will be added to the node palette under the **dashboard** category (*Figure 4.17*):

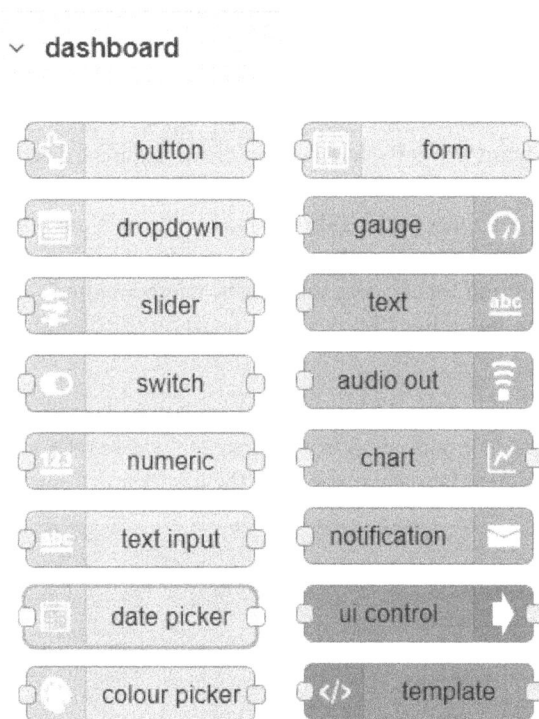

Figure 4.17 – dashboard nodes in the node palette

Moreover, there will be a new **dashboard** section in the control panel. This section will let us manage the layout, theme, or site of our dashboard. We will discuss these in detail now:

When you open the **dashboard** pane on the control panel from the tabs located on the left side of the screen, you will see the options shown in *Figure 4.18*:

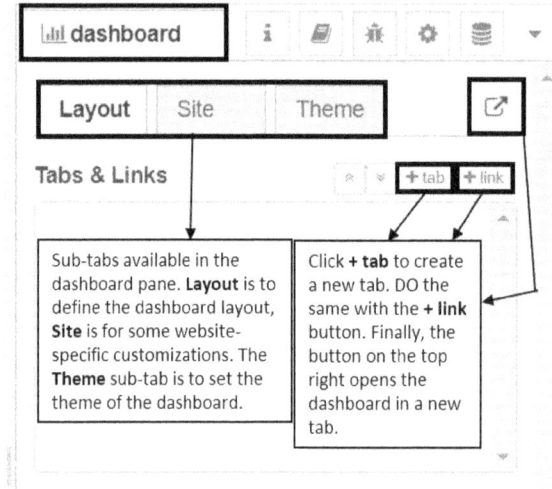

Figure 4.18 – The dashboard pane

By default, we are in the **Layout** sub-tab. Here, we can create different dashboard tabs and add all the UI elements to the desired tab according to our requirements. To add a new tab, just click the + **tab** button and a new tab will appear in the **Tabs & Links** window. Now, whenever you add a new UI node, you can select which tab (or group, to be more specific) it belongs to. You can even see the same element in this window and we can drag it from one tab to another if required. Different tabs can be accessed through the menu bar (the three-bar icon) on the dashboard (we've created two dummy tabs for the demo).

To demonstrate the dashboard in action, I have added two switch nodes from the **Dashboard** section and added them to two different groups, each in a separate tab. Please refer to *Figure 4.19* to see how you will be able to access different tabs when you open the dashboard:

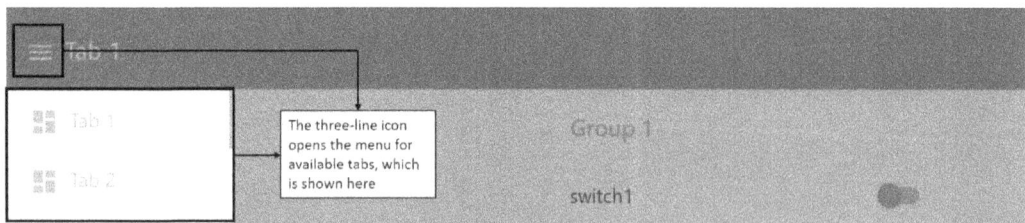

Figure 4.19 – How to switch tabs in the Node-RED dashboard

There are several other customizations in nodes, such as their size and other application-specific information. For example, consider a *Gauge* widget, which can be used to graphically display any incoming value. We can select the heading of this widget, the range, the value format, and even the unit that will be shown on the dashboard. Hence, we have highly customizable widgets, which help us build an interactive dashboard.

This concludes this section. In the next section, we will be utilizing all the knowledge we have learned so far to create a mini-project where we will be controlling the onboard NodeMCU LED using a switch widget on our Node-RED dashboard.

Mini project 2 – Controlling a NodeMCU LED from the Node-RED dashboard

The project's aim is to control the onboard NodeMCU LED using a switch widget on our Node-RED dashboard, where we will be using MQTT as our primary communication protocol. We will proceed step by step, covering the hardware requirements, software setup, code explanation, and project demonstration.

Let's first start with the hardware requirements for this project.

Hardware requirements

This is a very simple project so the hardware requirements are quite straightforward:

- A NodeMCU development board
- A Raspberry Pi

Information about both these devices has been covered in detail in the previous chapters. In the later sections, device-specific instructions that need to be followed will be listed.

Software requirements

As for the software requirements, only the Raspberry Pi needs to be considered as the NodeMCU code explanation will come later.

The following should have been installed, set up, and configured in the Pi:

- Raspberry Pi OS setup (refer to *Chapter 1, Introduction to Raspberry Pi and MQTT*)
- Mosquito MQTT package installation and setup (refer to *Chapter 1, Introduction to Raspberry Pi and MQTT*)
- Node-RED installation and configuration (covered in the earlier sections of this chapter)

Please note that all these steps have already been performed if you have been following the book. If any of these steps are left to do, please refer to the chapters mentioned previously to complete them.

Hence, we are good with the software setup too. Next, we will move on to the NodeMCU- and Raspberry Pi-specific setups.

NodeMCU setup

As far as NodeMCU is concerned, no external sensors or actuators need to be connected for this project. Hence, we just need to write and upload the code specific to this project.

We will now start with the code walk-through and explanation.

Code explanation

The code for this project is pretty straightforward. This code does the following:

- Connects to an MQTT server (the Pi's server in this case)
- Subscribes to the `project2/led` topic
- Changes the on-chip LED state according to the value received on the preceding topic
- Reconnects to the MQTT server if it disconnects

Now, we will cover the code in parts for better understanding. If you notice, the code is quite similar to the code written in *Chapter 2, MQTT in Detail*. The reason for that is the MQTT connection from the NodeMCU side remains the same. We are changing how we interact with our node (we used the terminal in Mini-Project 1 and now we will be using the Node-RED Dashboard UI). So, let's start with the code explanation.

Importing the required libraries:

```
#include <ESP8266WiFi.h>
#include <PubSubClient.h>
```

The first two lines of the code import the required libraries. The `ESP8266WiFi` library is used to access Wi-Fi networks, which grants the board internet access. The `Pubsubclient` library is the MQTT client library, which helps us run an MQTT client on the development board:

Important credentials and variable declarations:

```
const char* ssid = "wifi_ssid";
const char* password = "wifi_password";
const char* mqtt_server = "pi_ip_address";
```

```
WiFiClient espClient;
PubSubClient client(espClient);
unsigned long lastMsg = 0;
#define MSG_BUFFER_SIZE  (50)
char msg[MSG_BUFFER_SIZE];
int value = 0;
```

The next three lines are constant variable initializations for the *Wi-Fi name, password*, and the *MQTT (pi) IP address*. The Raspberry Pi's IP address is the IP address of the broker.

The next few lines deal with the various object and variable initializations required later in our code:

Wi-Fi setup function:

```
void setup_wifi()
{
  delay(10);

  Serial.println();
  Serial.print("Connecting to ");
  Serial.println(ssid);
  WiFi.mode(WIFI_STA);
  WiFi.begin(ssid, password);
  while (WiFi.status() != WL_CONNECTED)
  {
    delay(500);
    Serial.print(".");
  }

  randomSeed(micros());

  Serial.println("");
  Serial.println("WiFi connected");
  Serial.println("IP address: ");
  Serial.println(WiFi.localIP());
}
```

The setup_wifi function is used to connect to the Wi-Fi network whose credentials we provided as SSID and password constant variables.

Callback function for handling MQTT subscribed topics:

```
void callback(char* topic, byte* payload, unsigned int length)
{
  Serial.print("Message arrived [");
  Serial.print(topic);
  Serial.print("] ");
  for (int i = 0; i < length; i++) {
    Serial.print((char)payload[i]);
  }
  Serial.println();
// LED On Off Logic
  if ((char)payload[0] == '1') {
    digitalWrite(BUILTIN_LED, LOW);
  }
  else
  {
    digitalWrite(BUILTIN_LED, HIGH);
  }
}
```

The `callback` function is used to print out the data received on the subscribed MQTT channels. In this case, we are subscribed to the `project2/led` topic, to which we will be sending data to control the onboard LED. We have programmed the function such that if it receives `1` on the topic, the LED turns on, and if it receives zero on the topic, it turns off:

MQTT reconnect function:

```
void reconnect() {
  // Loop until we're reconnected
  while (!client.connected()) {
    Serial.print("Attempting MQTT connection...");
    // Create a random client ID
    String clientId = "ESP8266Client-";
    clientId += String(random(0xffff), HEX);
    // Attempt to connect
    if (client.connect(clientId.c_str())) {
      Serial.println("connected");
      // resubscribe to the specific topic
```

```
    client.subscribe("project2/led");
  } else {
    Serial.print("failed, rc=");
    Serial.print(client.state());
    Serial.println(" try again in 5 seconds");
    // Wait 5 seconds before retrying
    delay(5000);
  }
}
}
```

The reconnect function is used to reconnect to the MQTT broker if there is a problem and the board disconnects. The reasons for this may be internet connection failure, duplicate client IDs, and so on.

This function tries to connect to the MQTT broker every 5 seconds.

Arduino code's setup function:

```
void setup() {
  // Initialize the BUILTIN_LED pin as an output
  pinMode(BUILTIN_LED, OUTPUT);
  // Turn off the LED initially.
  digitalWrite(BUILTIN_LED, HIGH);
  // Serial port opened
  Serial.begin(115200);
  // Call the setup_wifi function
  setup_wifi();
  // Connect to the MQTT Server
  client.setServer(mqtt_server, 1883);
  // Define the Callback Function
  client.setCallback(callback);
  // Subscribe to this topic
  client.subscribe("project2/led");
}
```

This is the Arduino setup function. We set the built-in LED pin mode to output. Then, we initiate a serial connection with a baud rate of 115200. Next, we call the setup_wifi function to connect to the Wi-Fi network.

Next, we connect to the MQTT server, which in our case is the Raspberry Pi, and then we specify the `callback` function using the `setCallback` function of the `Pubsubclient` library:

Arduino Code's Loop Function

```
void loop() {
  if (!client.connected()) {
    reconnect();
  }
  client.loop();
}
```

The final `loop` function is the one that runs indefinitely. First, we check whether the MQTT client is disconnected. If so, we run the `reconnect` function.

In addition to this, we resubscribe to the `project2/led` topic so if we send a 0 or 1, we can control the LED on NodeMCU.

This completes the NodeMCU part of the project. Next, we will cover the Raspberry Pi setup.

Raspberry Pi setup

For this project, it is assumed that you have already set up your Raspberry Pi. That includes the installation of OS, the MQTT package, and also Node-RED (with the `node-red-dashboard` module installed).

We have to do the following tasks for the Pi:

1. Create a dashboard layout (this includes the creation of a tab and group) for the project.
2. Create the project flow and deploy these changes.

So, let's get started with the dashboard layout setup.

Dashboard layout setup

In this section, we will be creating a simple dashboard for this project on Node-RED. It will consist of a single tab with a switch widget, which will be used to control the NodeMCU onboard LED. To do this, just follow the step-by-step tutorial listed next:

1. Open Node-RED on your browser of choice. Then navigate and open the **dashboard** section from the control panel. The control panel is situated on the right side of the screen. Take a look at *Figure 4.20* for reference:

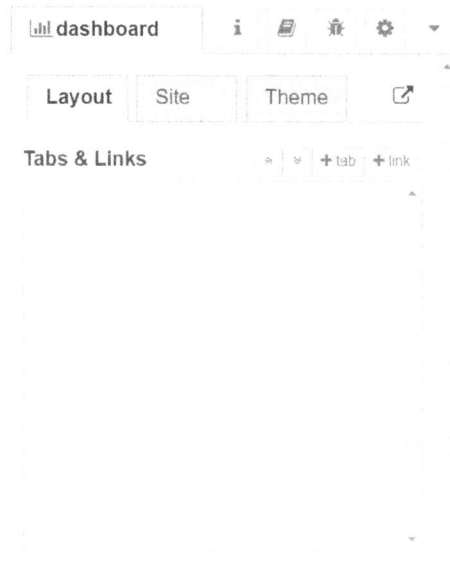

Figure 4.20 – The dashboard section in the Node-RED control panel

2. Next, click on the + **tab** button and then click on the **edit** button, which will appear when you hover your mouse pointer over the tab. In the edit window, rename the tab to what you want (I have simply named it **Project 2**) and then click the **Update** button. Refer to *Figure 4.21* for reference:

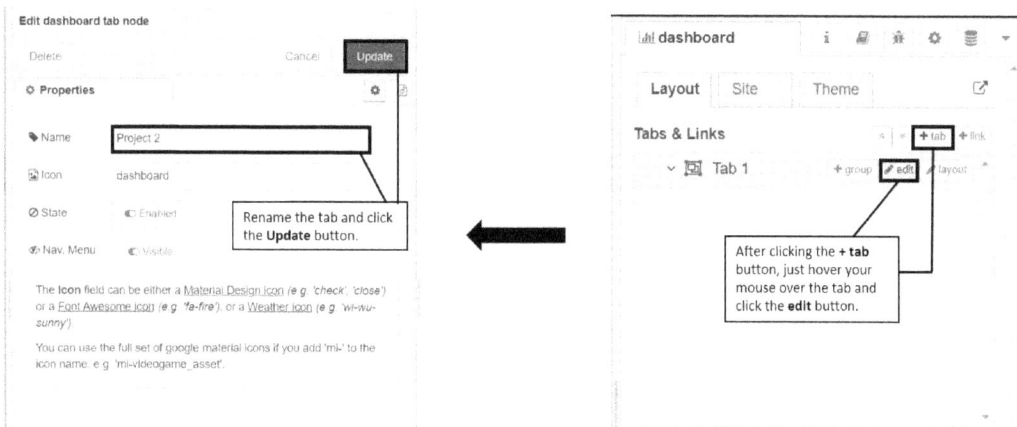

Figure 4.21 – Creating a new tab and renaming it

3. Next, we need to create a group in the tab as all the widgets we use in our flow need to be assigned to a particular group. We will create a new group using the **+ group** option, which appears when you hover over the tab.

4. Just clicking it will create a new group, which needs to be renamed as well, using the same process as for the tab. I have named the group **NodeMCU**. See *Figure 4.22* for reference:

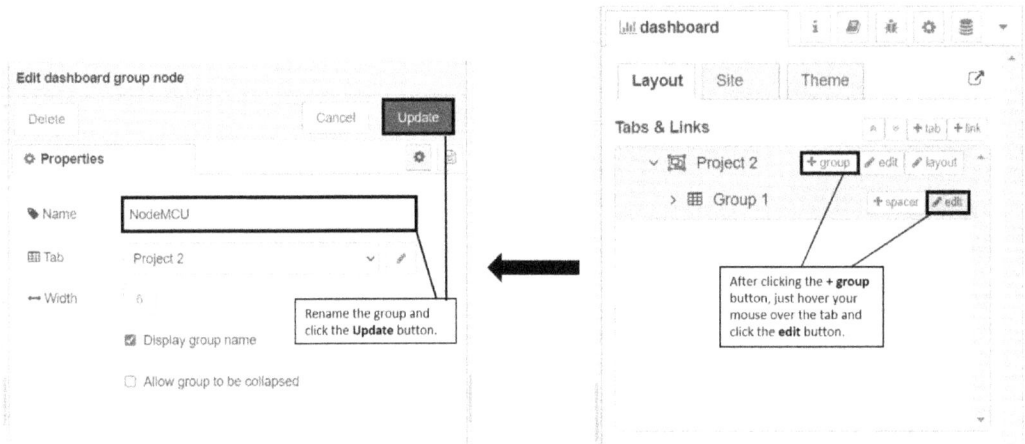

Figure 4.22 – Adding a new group to the tab and renaming it

This completes the dashboard layout setup. We only require a single group for this project as we will be adding only a single widget.

Next, we will create the flow for this project.

Project flow

We will be creating a simple flow that performs the following task:

It will create a switch widget on your dashboard, which will send 0 and 1 on and off positions respectively, and this message will then be published on the project2/led topic as payload.

Let's start creating the flow. Just follow these steps:

1. Drag two nodes from the node palette. The node names are as follows:

    -   The **switch** node under the **dashboard** section

    -   The **mqtt out** node under the **network** section

Please refer to *Figure 4.23* for reference:

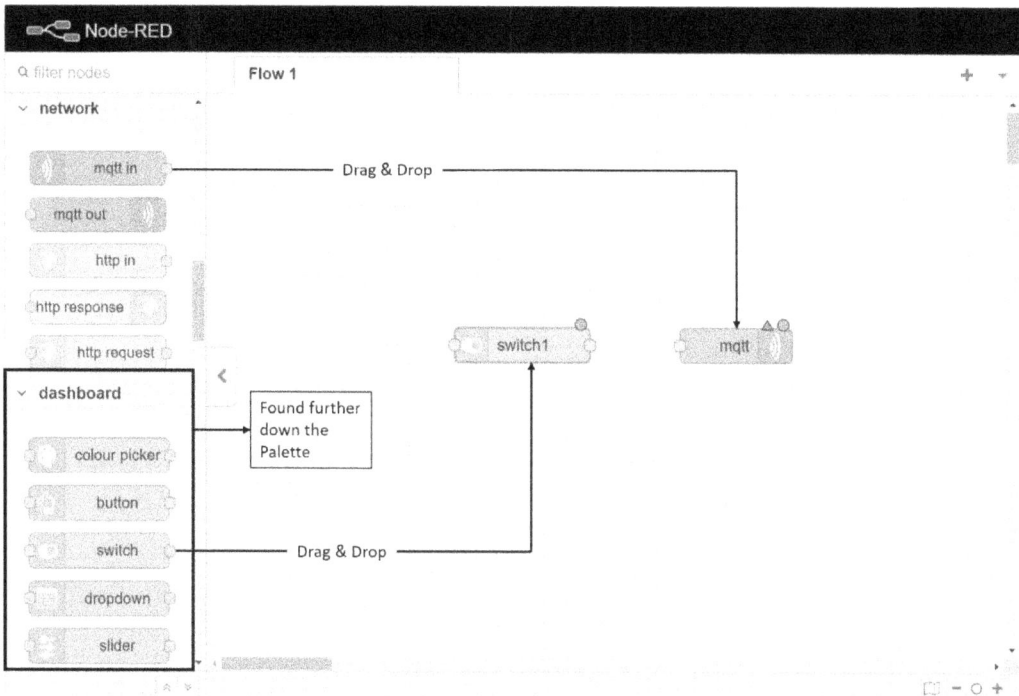

Figure 4.23 – Drag and drop the project flow components onto the workspace

2. Connect the switch node to the **mqtt out** node by dragging a wire across both. We will now have to configure the nodes for our project. Please note that the following customizations for each of the two nodes will be required to complete the flow:

- The **switch node** will require the *Group*, *Label*, *On Payload*, *Off Payload*, and *Name* sections to be changed.

- The **mqtt out** node will require the *Server* and *Name* to be set up.

Please refer to *Figure 4.24* to complete the listed customizations and, hence, the flow for this particular project:

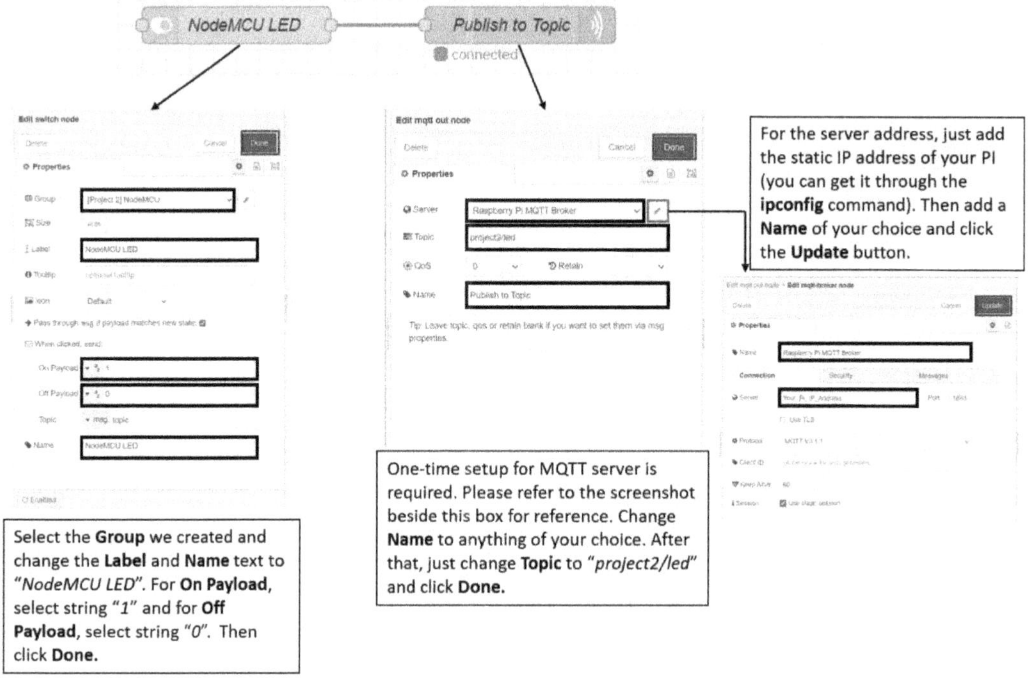

Figure 4.24 – Project flow setup for mini project 2

This completes the project flow setup for this project. Just deploy this flow by clicking the **Deploy** button situated on the top-right side of the screen. Now, to view our dashboard, just open the Node-RED dashboard using the link mentioned earlier in the chapter. Our dashboard will only have a single tab, and on it, we will find a single switch situated in the middle of the screen. Refer to *Figure 4.25* to see what our dashboard will look like:

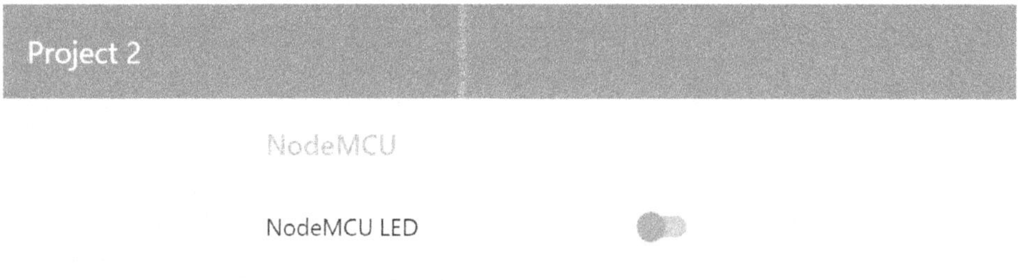

Figure 4.25 – Mini Project 2 Node-RED dashboard

You will now be able to control your NodeMCU's on-chip LED through the switch on the dashboard. This is just a small demonstration of what this platform can do. This is the end of this chapter's last section.

Summary

This chapter mainly focused on a single piece of software that we will be using a lot throughout this book – Node-RED. We started with an introduction to this software, followed by a tutorial on how to install/update it to the latest version on our Raspberry Pi. After that, we had a crash course on the basics of how to use it to its fullest extent. Next, we covered the MQTT components of Node-RED (the nodes specifically) and installed and got acquainted with the Node-RED dashboard. Finally, we wrapped up by creating a mini-project wherein we controlled the onboard NodeMCU board wirelessly through a simple dashboard hosted on the Raspberry Pi.

Now that we have covered all the building blocks for this book (Raspberry Pi, MQTT, and Node-RED), in the next chapter, we will be creating our first major project: a **weather station** based on NodeMCU. This project will have a far more complex and interactive dashboard and it will show us just how powerful and easy to use Node-RED is.

Part 2: Practical Implementation – Building Two Full-Scale Projects

Now that we possess all the fundamental knowledge about Raspberry Pi, MQTT, ESP development boards, and Node-RED, it is time to implement this knowledge by developing two full-scale projects!

This part comprises of the following chapters:

- *Chapter 5, Major Project 1: IoT Weather Station*
- *Chapter 6, Major Project 2: Smart Home Control Relay System*

5
Major Project 1: IoT Weather Station

Now that we are more knowledgeable about the topics discussed in the previous chapters, we will be making our first major project: an *IoT weather station*. This chapter gives step-by-step instructions on how to build this. The instructions will be divided into the following sections:

- Hardware requirements
- Code explanation
- Raspberry Pi setup

The aim of this project is to build a fully functional weather station (based on the popular NodeMCU development board) whose readings can be monitored on a Node-RED dashboard in real time, which will be hosted on the Raspberry Pi. Note that we will be using MQTT as the communication protocol between the NodeMCU and the Raspberry Pi, whose host is also the Pi (hence, the dashboard will be only available on the local network). The final breadboard circuit is shown in *Figure 5.1*.

Figure 5.1 – Your very own NodeMCU-based weather station!

We will now look at the hardware requirements before moving on to building the weather station.

Hardware requirements

To build our weather station, we will require a development board that fetches the sensor values and sends them to a particular destination, Node-RED in our case, through a communication protocol. For the development board, we will utilize the NodeMCU development board for this project primarily due to its relevant features and cost-effectiveness. Please refer to the following figure to find the components required to build our weather station (*Figure 5.2*).

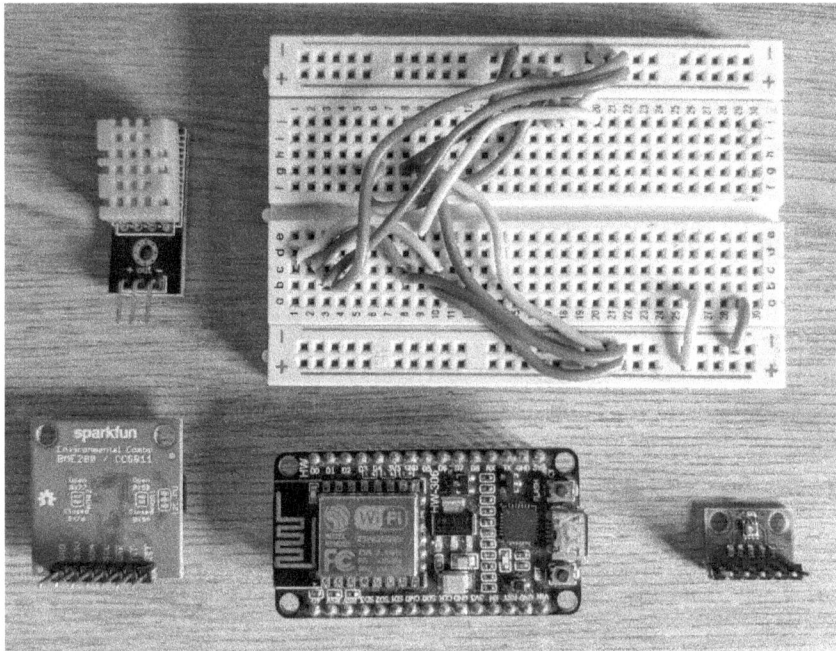

Figure 5.2 – The required hardware for the project

Next, we need to choose appropriate sensors so that the readings we get are both reliable and accurate. Hence, we choose the following three sensors for our project:

- A DHT11 temperature and humidity sensor
- A BMP280 pressure sensor
- A CCS811 air quality sensor

We will also need something that we can use to interface all the sensors to the development board. For this, we will use a breadboard and some connecting wires. Now, we will briefly discuss each component that we will be using in this project.

The NodeMCU development board

The NodeMCU is an immensely popular development board based on the ESP8266 chip. The most important feature of this board is its ability to connect and use Wi-Fi for communication. Besides this, it has a lot of GPIO pins and supports I2C, SPI, and PWM. Therefore, its features paired with its cheap price make it the perfect choice for numerous IoT projects.

A NodeMCU development board is shown in *Figure 5.3*.

Figure 5.3 – A NodeMCU development board

The general features of this board are as follows:

- It is easy to use
- It can be configured to act as an access point or station
- It can be used for event-driven API applications
- No external antenna is required for Wi-Fi connection (internal antenna provided)
- It contains 13 GPIO pins, 10 PWM channels, I2C, SPI, ADC, UART, and 1-Wire
- It can be programmed using the popular open source IDE platform, Arduino IDE

The DHT11 temperature and humidity sensor

The DHT11 sensor can be used to measure temperature and humidity values and communicate them serially over a single wire. This is a fairly basic sensor with intermediate accuracy. The value range for each is as follows:

- Temperature: 0 to 50°C
- Humidity: relative humidity given as a percentage (20 to 90%)

Have a look at what the commercially available DHT11 sensor looks like in *Figure 5.4*.

Figure 5.4 – A DHT11 temperature and humidity sensor (three-pin)

The sensor can have three or four pins. They are *VCC*, *GND*, *NC* (this pin is not used and hence is not included in the three-pin format), and *Signal*. The sensor can be powered by 3.3 V or 5 V. We need to connect the sensor **GND** (**Ground**) to the development board GND and the Signal pin needs to be connected to a digital pin.

The BMP280 temperature and pressure sensor

The BMP280 is a sensor that can be used to measure barometric pressure and altitude readings. In addition to this, it also gives out temperature readings, and this data is accessible via an SPI or I2C communication protocol.

Figure 5.5 shows what a commercially available BMP280 sensor looks like.

Figure 5.5 – A BMP280 temperature and pressure sensor

We will only use the pressure and altitude values for this project, and we will use the **Inter-Integrated Circuit (I2C)** communication protocol. The characteristics of this sensor are listed here:

- The sensor's operating voltage range is 1.71 to 3.6 volts.

- The sensor's operating temperature range is between -40 and 85 degrees Celsius but it provides the most accurate measurements between 0 and 65 degrees.

- The sensor has a peak current threshold of 1.12 mA.

- The operating pressure ranges between 300 hPa and 1100 hPa.

The CCS811 air quality sensor

The CCS811 is a low-cost air quality sensor that has the capability of measuring the **volatile organic compounds (VOCs)** in an indoor environment using a metal oxide gas sensor. Additionally, it also has the capability to output the **equivalent CO2 (eCO2)** values.

In terms of the hardware itself, it supports both an **analog-to-digital converter (ADC)** and I2C interface. Moreover, it supports a number of drive modes, which help us configure the interval between two consecutive readings. This is a very important feature, as it helps us optimize the overall power consumption during active measurement cycles, giving the device extended battery life, especially for portable devices.

Please refer to the following figure (*Figure 5.6*) to see what the actual sensor looks like.

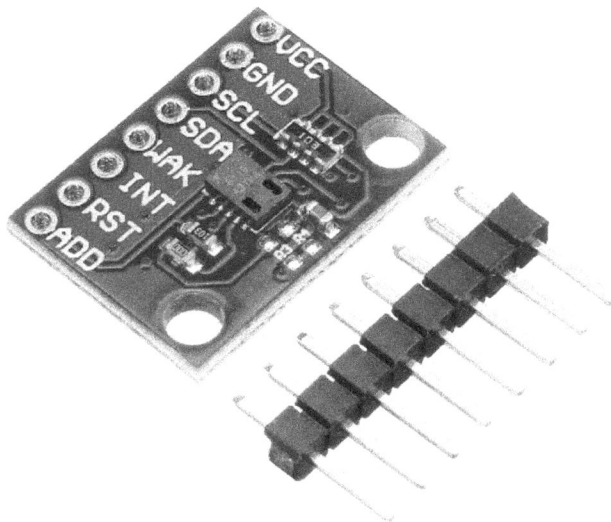

Figure 5.6 – A CCS811 air quality sensor

The CCS811 has five modes of operation:

- **Mode 0**: Idle, low-current mode

- **Mode 1**: Constant power mode, with IAQ measurement every second

- **Mode 2**: Pulse heating mode, with IAQ measurement every 10 seconds

- **Mode 3**: Low-power pulse heating mode, with IAQ measurement every 60 seconds

- **Mode 4**: Constant-power mode, with sensor measurement every 250 milliseconds

These are all the required components. The next section will cover how to interface them.

Sensor interfacing

Now that we have all the required components, we just need to connect everything together. For that, we will be using a breadboard and connecting wires. Please note that the CCS811 and BMP280 sensors will be connected to the NodeMCU board via I2C and that the DHT11 sensor will be connected to the digital pin for sensor value transmission. A schematic diagram that represents this is shown in *Figure 5.7*.

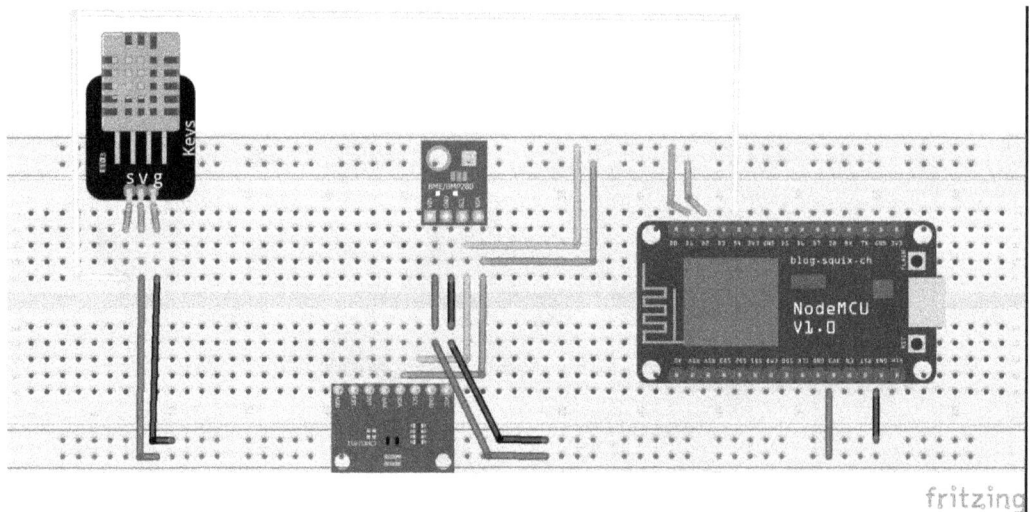

Figure 5.7 – The hardware sensor interfacing for Major Project 1

The circuit schematic for this project is pretty straightforward. The NodeMCU's D1 and D2 pins are multiplexed and can also be used as I2C pins (SDA and SCL). The DHT11 sensor is connected to digital pin D4.

After the above connections have been made, we are ready to move on to the next step of our project, which is writing the code for the weather station and flashing it into your NodeMCU development board.

Code explanation

The hardware has been set up and now we need to write and flash the code for it. The code will do the following:

1. Configure the pins for connecting the sensors.

2. Connect to the MQTT broker (the Pi's broker in this case).

3. Subscribe to the relevant MQTT topics.

4. Get the sensor values and publish them to their particular topics. This will reflect on the dashboard in real time.

5. Reconnect to the MQTT server if it disconnects.

The last two steps will run indefinitely. Now, we will look at the code in chunks, and finally, we will look at the whole code and the relevant GitHub link. So, let's get started with the code explanation. As the code is a little complicated, the code has been divided into important subsections for clarity and better understanding.

To import the required libraries, we need the following:

```
#include <ESP8266WiFi.h>
#include <PubSubClient.h>
#include <Wire.h>
#include <Adafruit_BMP280.h>
#include "SparkFunCCS811.h"
#include "DHT.h"
```

First, we will need to import all the required libraries. That includes specific libraries for the BMP280, CCS811, and DHT sensors. These are specific Arduino libraries written to help you easily get readings from these sensors. To know more about them, just follow the links below to the GitHub repository for each library:

*   Adafruit_BMP280 – https://github.com/adafruit/Adafruit_BMP280_Library

*   CCS811 – https://github.com/sparkfun/SparkFun_CCS811_Arduino_Library

*   DHT11 – http://www.github.com/markruys/arduino-DHT

Equally, the Wire library corresponds to I2C connections, the ESP8266WiFi library enables the Wi-Fi capability, and the Pubsubclient library connects to the MQTT brokers.

To define constants, variables, and objects, we need the following:

```
// Constants
#define CCS811_ADDR 0x5B

// Variables
const char* ssid = "wifi_ssid";
const char* password = "wifi_password";
const char* mqtt_server = "pi_ip_address";

// Objects
WiFiClient espClient;
PubSubClient client(espClient);
Adafruit_BMP280 bmp; // I2C
CCS811 mySensor(CCS811_ADDR);
DHT dht;
```

Next, we have to define the constants, variables, and objects for the project. These include the *Wi-Fi and MQTT credentials*, *objects for various sensor library classes*, and *Wi-Fi and MQTT client initialization*.

For the Wi-Fi setup function, see the following:

```
// Custom function for Wifi connection establishment
void setup_wifi()
{
  delay(10);
  Serial.println();
  Serial.print("Connecting to ");
  Serial.println(ssid);
  WiFi.mode(WIFI_STA);
  WiFi.begin(ssid, password);
  while (WiFi.status() != WL_CONNECTED)
  {
    delay(500);
    Serial.print(".");
  }
  randomSeed(micros());
  Serial.println("");
```

```
  Serial.println("WiFi connected");
  Serial.println("IP address: ");
  Serial.println(WiFi.localIP());
}
```

The setup_wifi function is used to connect to the Wi-Fi network, the credentials for which we provided as ssid and password constant variables.

For the MQTT Callback Function, see the following:

```
// Callback Function
void callback(char* topic, byte* payload, unsigned int length)
{
  Serial.print("Message arrived [");
  Serial.print(topic);
  Serial.print("] ");
  for (int i = 0; i < length; i++) {
    Serial.print((char)payload[i]);
  }
  Serial.println();
}
```

The callback function is used to print the data received on the subscribed MQTT channels. Additionally, we can write code that performs specified actions based on the messages received on particular topics.

As this project mostly deals with capturing the sensor data and transmitting it (that is, publishing it) to our dashboard, there will be no additional logic included in the callback function. We will cover this concept in detail in *Chapter 6, Major Project 2: Smart Home Control Relay System*, where we will create an ESP32-based smart-switching system, where we will control the switches wirelessly through a Node-RED dashboard. There, we will be subscribing to specific topics and controlling the state of a relay based on the message we receive on that particular topic.

For the MQTT reconnect function, see the following:

```
// Function to reconnect to MQTT server
void reconnect() {
  // Loop until we're reconnected
  while (!client.connected()) {
    Serial.print("Attempting MQTT connection...");
    // Create a random client ID
```

```
        String clientId = "ESP8266Client-";
        clientId += String(random(0xffff), HEX);
        // Attempt to connect
        if (client.connect(clientId.c_str()))
        {
          Serial.println("connected");
  // resubscribe to the specific topic
          client.subscribe("IoTWeatherStation/temperature/celcius");
          client.subscribe("IoTWeatherStation/temperature/
farenhiet");
          client.subscribe("IoTWeatherStation/humidity");
          client.subscribe("IoTWeatherStation/pressure");
          client.subscribe("IoTWeatherStation/altitude");
          client.subscribe("IoTWeatherStation/TVOC");
          client.subscribe("IoTWeatherStation/eCO2");
          client.subscribe("IoTWeatherStation/hic");
        }
        else {
          Serial.print("failed, rc=");
          Serial.print(client.state());
          Serial.println(" try again in 5 seconds");
          // Wait 5 seconds before retrying
          delay(5000);
        }
      }
    }
```

The MQTT reconnect function is in place to establish a connection to the broker again, in case there are issues from the NodeMCU side. These issues can be anything from internet failure to hardware problems.

This function tries to reconnect to the broker every five seconds and once the connection has been re-established, it resubscribes to all the important MQTT topics.

For computing the heat index, which is optional, see the following:

```
  float computeHeatIndex(float temperature, float
  percentHumidity) {
    float hi;
    temperature = 1.8*temperature+32; //convertion to *F
```

```
  hi = 0.5 * (temperature + 61.0 + ((temperature - 68.0) * 1.2)
+ (percentHumidity * 0.094));
  if (hi > 79) {
    hi = -42.379 +
              2.04901523 * temperature +
             10.14333127 * percentHumidity +
             -0.22475541 * temperature*percentHumidity +
             -0.00683783 * pow(temperature, 2) +
             -0.05481717 * pow(percentHumidity, 2) +
              0.00122874 * pow(temperature, 2) * percentHumidity
+
              0.00085282 * temperature*pow(percentHumidity, 2) +
             -0.00000199 * pow(temperature, 2) *
              pow(percentHumidity, 2);

    if((percentHumidity < 13) && (temperature >= 80.0) &&
(temperature <= 112.0))
       hi -= ((13.0 - percentHumidity) * 0.25) * sqrt((17.0 -
abs(temperature - 95.0)) * 0.05882);

    else if((percentHumidity > 85.0) && (temperature >= 80.0)
&& (temperature <= 87.0))
       hi += ((percentHumidity - 85.0) * 0.1) * ((87.0 -
temperature) * 0.2);
  }

  hi = (hi-32)/1.8;
  return hi; //return Heat Index, in *C
}
```

The `computeHeatIndex` function is used to calculate the heat index value from the temperature and humidity values. Please note that if you use the `DHT` library from `Adafruit` instead of the one used here, this function is available in the library itself.

For the `setup()` function, see the following:

```
void setup()
{
  Serial.begin(115200);
  Wire.begin(); //Initialize I2C Hardware
```

```
   dht.setup(D4);

   if (mySensor.begin() == false)
   {
     Serial.print("CCS811 error. Please check wiring.
Freezing...");
     while(1);
   }

   if (!bmp.begin(0x76)) {
     Serial.println(F("Could not find a valid BMP280 sensor,
check wiring or try a different address!"));
     while(1) { delay(10); }
   }

   // Default settings from datasheet
   bmp.setSampling(Adafruit_BMP280::MODE_NORMAL,
                   Adafruit_BMP280::SAMPLING_X2,
                   Adafruit_BMP280::SAMPLING_X16,
                   Adafruit_BMP280::FILTER_X16,
                   Adafruit_BMP280::STANDBY_MS_500);

   setup_wifi();

   client.setServer(mqtt_server, 1883);
   client.setCallback(callback);

   client.subscribe("IoTWeatherStation/temperature/celcius");
   client.subscribe("IoTWeatherStation/temperature/farenhiet");
   client.subscribe("IoTWeatherStation/humidity");
   client.subscribe("IoTWeatherStation/pressure");
   client.subscribe("IoTWeatherStation/altitude");
   client.subscribe("IoTWeatherStation/TVOC");
   client.subscribe("IoTWeatherStation/eCO2");
   client.subscribe("IoTWeatherStation/hic");
}
```

The setup() function does the following:

- Opens a serial port connection with a baud rate of 115200
- Enables I2C connectivity with Wire.begin()
- Sets up the DHT sensor pin for the GPIO D4 on the NodeMCU
- Checks whether the BMP280 and CCS811 sensors are working properly
- Establishes the BMP280 sensor settings
- Connects to Wi-Fi using the setup_wifi function
- Connects to the MQTT broker and subscribes to all the necessary topics

To reconnect to the MQTT broker logic, see the following:

```
void loop()
{
  // Reconnect to MQTT Broker Logic
  if (!client.connected()) {
    reconnect();
  }
```

This code block checks for any active MQTT connections and if it does not find one, it runs the reconnect function.

For the Variable Initialization, see the following:

```
//Variable Initialization
float co2val;
float tvocval;
static char temperatureC[7];
static char temperatureF[7];
static char humid[7];
static char co2[7];
static char tvoc[7];
static char pressure[7];
static char altitude[7];
static char hic[7];
```

This code block initializes all the variables that will be used in this loop function.

For the `Sensor Value Assignment`, see the following:

```
// Sensor Value Assignment
if (mySensor.dataAvailable())
{
  mySensor.readAlgorithmResults();
  co2val = mySensor.getCO2();
  tvocval = mySensor.getTVOC();
}

float temperature_C = bmp.readTemperature();
float pressureval = bmp.readPressure();
float altitudeval = bmp.readAltitude(1013.25);

float humidity = dht.getHumidity();
float hi = computeHeatIndex(temperature_C, humidity);
float temperature_F = dht.toFahrenheit(dht.getTemperature());
delay(2000);
```

Now that we have all the variables, we will read the sensor values using the functions provided by the sensor libraries and store them in their respective variables.

To convert float values into string values, see the following:

```
// Convert Float values to String (in Character Array format)
  dtostrf(temperature_C, 6, 2, temperatureC);
  dtostrf(temperature_F, 6, 2, temperatureF);
  dtostrf(humidity, 6, 2, humid);
  dtostrf(co2val, 6, 2, co2);
  dtostrf(tvocval, 6, 2, tvoc);
  dtostrf(pressureval, 6, 2, pressure);
  dtostrf(altitudeval, 6, 2, altitude);
  dtostrf(hi, 6, 2, hic);
```

To publish these values through MQTT topics, we will need to convert these float or decimal values into string datatypes (specifically in char array format). Hence, we will use the `dtostrf` function to accomplish this.

To publish the sensor data, see the following:

```
// Publish the sensor data on their particular MQTT topics
client.publish("IoTWeatherStation/temperature/celcius",
temperatureC);
client.publish("IoTWeatherStation/temperature/farenhiet",
temperatureF);
client.publish("IoTWeatherStation/humidity", humid);
client.publish("IoTWeatherStation/pressure", pressure);
client.publish("IoTWeatherStation/altitude", altitude);
client.publish("IoTWeatherStation/TVOC", tvoc);
client.publish("IoTWeatherStation/eCO2", co2);
client.publish("IoTWeatherStation/hic", hic);
```

Next, we will publish all the sensor values to their specific MQTT topics. We will be using the Pubsubclient library's `client.publish` function to achieve this.

For printing the sensor values, see the following:

```
// Printing the Sensor Values on Serial Monitor
Serial.print
ln("-----------------------------------------------------------");
Serial.print("Temperature: ");
Serial.println(temperature_C);
Serial.print("Humidity: ");
Serial.println(humidity);
Serial.print("Heat Index Value: ");
Serial.println(hic);
Serial.print("TVOC Value: ");
Serial.println(tvocval);
Serial.print("eCO2 Value: ");
Serial.println(co2val);
Serial.print("Pressure Value: ");
Serial.println(pressureval);

Serial.print("Altitude Value: ");
Serial.println(altitudeval);
Serial.print
```

```
ln("-----------------------------------------------------------");
}
```

Finally, we will be printing the sensor values on the serial monitor so that when you see unusual sensor values on your dashboard, you know that debugging is required.

This completes the code explanation section. Please note that the whole code for this project is available in the GitHub repository created for this book. Please find the link for the same repository here:

`https://github.com/PacktPublishing/Raspberry-Pi-and-MQTT-Essentials`

In the next section, we will set up our Raspberry Pi for this particular project. This includes creating a new dashboard for this project and setting up a Node-RED flow.

Raspberry Pi setup

The Raspberry Pi will be the host for the local MQTT broker in this project and also the dashboard hosting device. The dashboard for this project will be created using Node-RED and the Node-RED dashboard module, which will both be running on the Pi.

The previous chapters cover in detail how to set up and activate the MQTT broker (refer to *Chapter 1, Introduction to Raspberry Pi and MQTT*, to refresh your memory). If you have followed the book by chapter, you will already have the MQTT broker up and running on boot. The next step is to create the Node-RED flow and the dashboard for our project. Please follow these step-by-step instructions, which will walk you through the entire setup process.

Starting Node-RED

Run the `node-red-start` command on the Pi to start Node-RED at the following IP address: `<pi's ip address>:1880`. Please refer to *Figure 5.8* for reference.

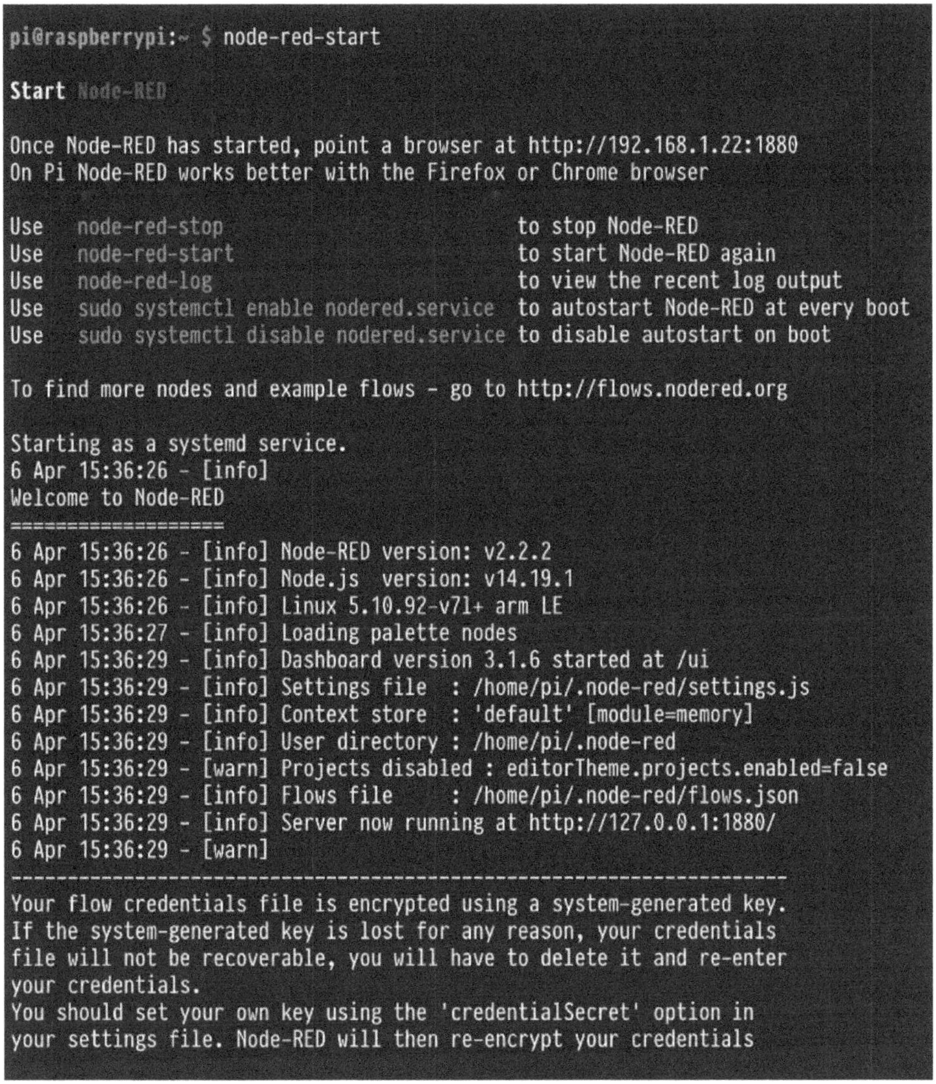

```
pi@raspberrypi:~ $ node-red-start

Start Node-RED

Once Node-RED has started, point a browser at http://192.168.1.22:1880
On Pi Node-RED works better with the Firefox or Chrome browser

Use    node-red-stop                          to stop Node-RED
Use    node-red-start                         to start Node-RED again
Use    node-red-log                           to view the recent log output
Use    sudo systemctl enable nodered.service  to autostart Node-RED at every boot
Use    sudo systemctl disable nodered.service to disable autostart on boot

To find more nodes and example flows - go to http://flows.nodered.org

Starting as a systemd service.
6 Apr 15:36:26 - [info]
Welcome to Node-RED
===================
6 Apr 15:36:26 - [info] Node-RED version: v2.2.2
6 Apr 15:36:26 - [info] Node.js  version: v14.19.1
6 Apr 15:36:26 - [info] Linux 5.10.92-v7l+ arm LE
6 Apr 15:36:27 - [info] Loading palette nodes
6 Apr 15:36:29 - [info] Dashboard version 3.1.6 started at /ui
6 Apr 15:36:29 - [info] Settings file  : /home/pi/.node-red/settings.js
6 Apr 15:36:29 - [info] Context store  : 'default' [module=memory]
6 Apr 15:36:29 - [info] User directory : /home/pi/.node-red
6 Apr 15:36:29 - [warn] Projects disabled : editorTheme.projects.enabled=false
6 Apr 15:36:29 - [info] Flows file     : /home/pi/.node-red/flows.json
6 Apr 15:36:29 - [info] Server now running at http://127.0.0.1:1880/
6 Apr 15:36:29 - [warn]
------------------------------------------------------------------------
Your flow credentials file is encrypted using a system-generated key.
If the system-generated key is lost for any reason, your credentials
file will not be recoverable, you will have to delete it and re-enter
your credentials.
You should set your own key using the 'credentialSecret' option in
your settings file. Node-RED will then re-encrypt your credentials
```

Figure 5.8 – Starting Node-RED on the Raspberry Pi through the terminal

We will now look at the setup of the dashboard.

Project flow and dashboard setup

Next, on the home screen of Node-RED, create a new flow following the instructions provided in *Figure 5.9* to get a new, blank workspace.

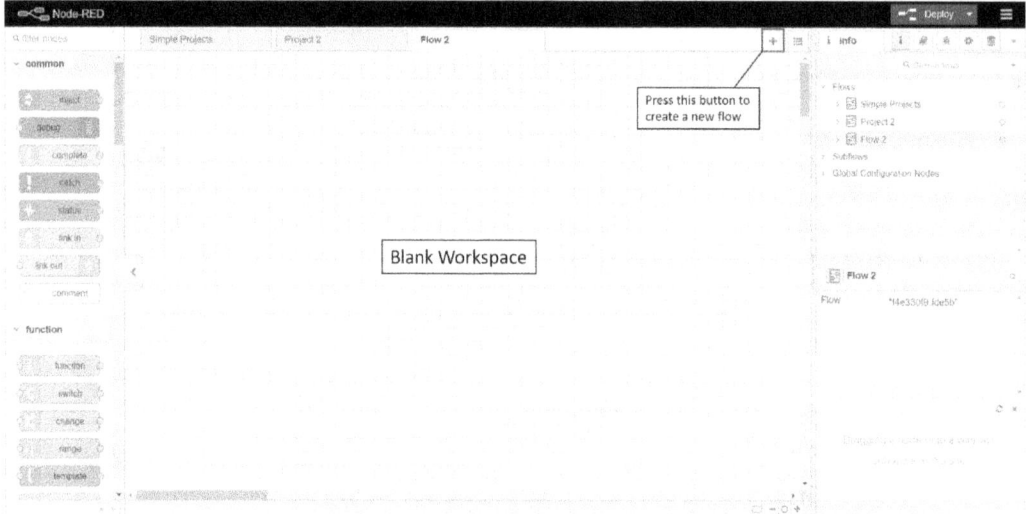

Figure 5.9 – Creating a new flow in Node-RED

From the node palette, just drag the following nodes into the blank workspace. We will put them all into the workspace first and then we will connect them together to create our flow, incorporating the following nodes:

- Eight mqtt in nodes (from the *network section*)
- Two text nodes (from the *dashboard section*)
- Two chart nodes (from the *dashboard section*)
- Four gauge nodes (from the *dashboard section*)

After dragging these nodes into the workspace, there will be 16 nodes in the blank space in total. Please have a look at *Figure 5.10* for reference.

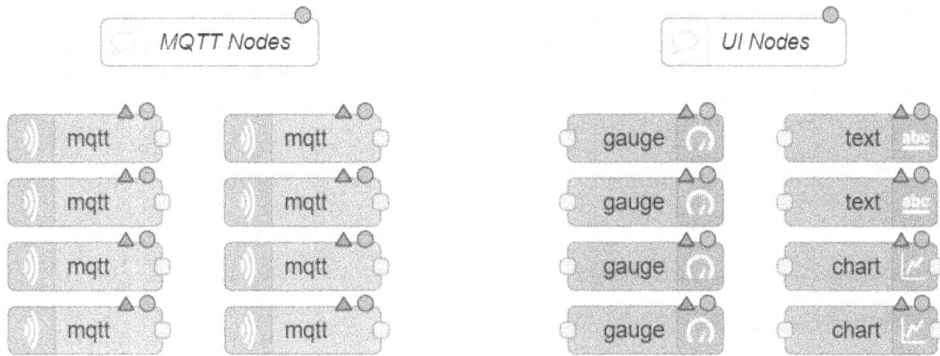

Figure 5.10 – Nodes to be dragged into the workspace

Before moving on to the configuration of nodes, we need to create a layout for the dashboard. We will choose a 2x4 layout, which simply means that there will be two rows on the dashboard with four widgets in each row.

For this, we need to create a new tab for our project and then create four different groups (with groups acting as columns). Each group will have two widgets.

Please follow *Figure 5.11* to refer to what the layout and the dashboard layout setup will look like.

Dashboard Layout Setup
❖ First, create a new Tab using the **+tab** button and rename to *"IoT Weather Station"*.
❖ Then using the **+group** button on the tab, create four groups as shown in the figure.
❖ Now, all you have to do is add your widgets to individual groups. For this project, we will follow the layout shown below. Hence, we will be adding a *Gauge* and any of the other two widgets in each group.

IoT Weather Station Dashboard Layout

Figure 5.11 – The Node-RED dashboard layout for this project

Now that we have all the nodes in our workspace, we will set up each node. That includes all the UI nodes being set up according to the value they will handle and the MQTT In nodes being set up for different topics. Please note that this tutorial assumes that you have already configured your Pi's MQTT broker in Node-RED. If you haven't, please follow the *Project* part of *Chapter 4, Node-RED on Raspberry Pi*.

Follow the preceding figure to group the UI nodes accordingly. For instance, in this project, we have chosen to show the following values:

1. **temperature** (*Celsius*), **temperature** (*Fahrenheit*), **humidity**, and **altitude** as gauges

2. **pressure** and **heat index** values as text displays

3. **TVOC** and **eCO2** values as charts

Hence, we will be grouping the nodes accordingly. There are eight **mqtt in** nodes and four **gauge** nodes. Please follow the instructions to fill in all the required details. Please refer to *Figure 5.12* to understand how to set up the mqtt in nodes and the gauge nodes.

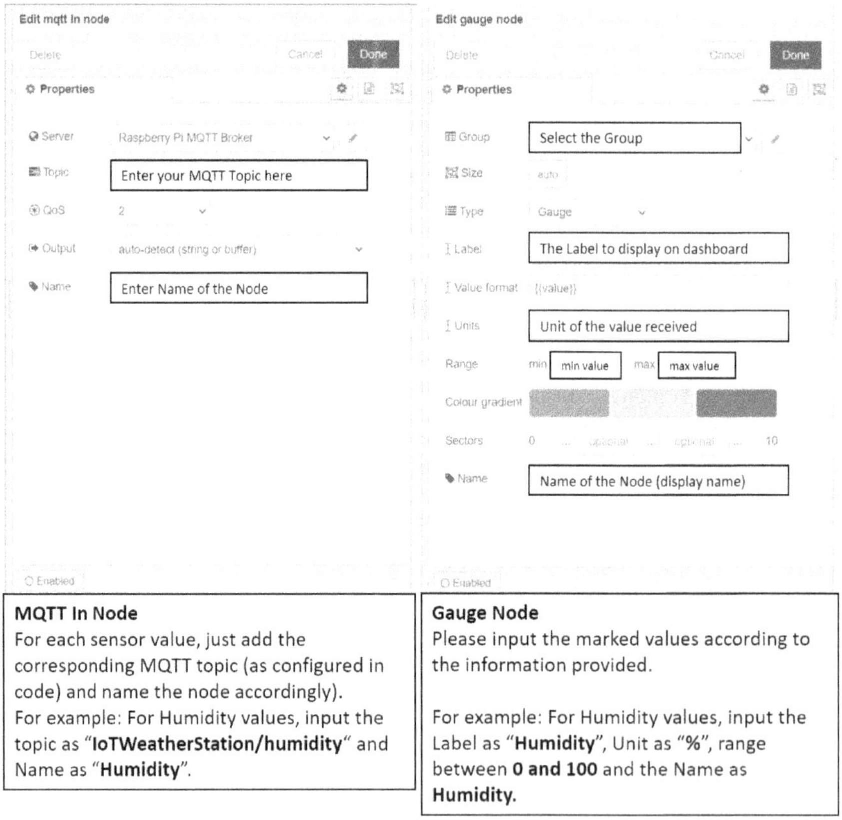

MQTT In Node	Gauge Node
For each sensor value, just add the corresponding MQTT topic (as configured in code) and name the node accordingly). For example: For Humidity values, input the topic as "**IoTWeatherStation/humidity**" and Name as "**Humidity**".	Please input the marked values according to the information provided. For example: For Humidity values, input the Label as "**Humidity**", Unit as "**%**", range between **0 and 100** and the Name as **Humidity**.

Figure 5.12 – mqtt in and gauge node configuration for this project

Important note

Please bear in mind that the topic names and the node names mentioned in the preceding figure have been chosen according to the code and personal preference. The code explanation contains all the topic names used for this project. Please name the topics and nodes in a meaningful way so that they are easy to understand.

Once the configuration for these nodes is complete, we will move on to the `text` nodes and the `chart` nodes. The `text` node shows a single value with a label that changes in real time according to the received data. On the other hand, the `chart` node shows a chart (in our case, a line chart) where the x-axis is the time axis, and the y-axis is the sensor value. We will display the pressure and heat index values in this format.

For this project, we will choose to display all the values within a two-minute window. As for which values to use this widget for, we have chosen the air quality sensor values (TVOC and eCO2). Please refer to *Figure 5.13* to complete the configuration for the `text` nodes and the `chart` nodes.

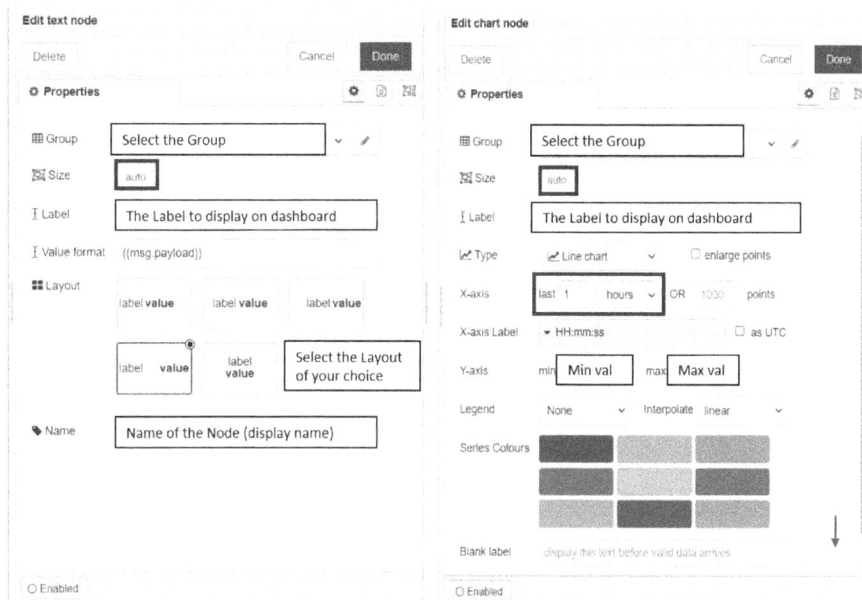

Text Node	**Chart Node**
Please fill in the marked values according to the information provided. Set size to **6x6**.	We will create line charts. Fill in the text box info, set size to **6x6**, and for x-axis, just enter a **time range** (2 minutes). Then add the **range** of sensor values (Eg: 0 - 50°C). Lastly, the down arrow indicates that the screen needs to scrolled. The last thing to input is the *Name* of the node.
For example: For Pressure values, input *Label* as "**Pressure**", add the unit in *Value format* as "**{{msg.payload}} Pa**" and input "**Pressure**" in *Name*.	

Figure 5.13 – Text and chart node configuration

We will also increase the font of the label. To increase the font, type `<font size = 12> Label_name`.

This completes the configuration of all our nodes for the flow. The next step is to connect all the nodes to complete the flow. We will have to connect each `mqtt in` component to its corresponding `dashboard` component (refer to *Figure 5.12*) so that whatever incoming data arrives on a particular MQTT topic will be reflected on our dashboard in real time.

Please refer to *Figure 5.14* for assistance on how to connect the project flow.

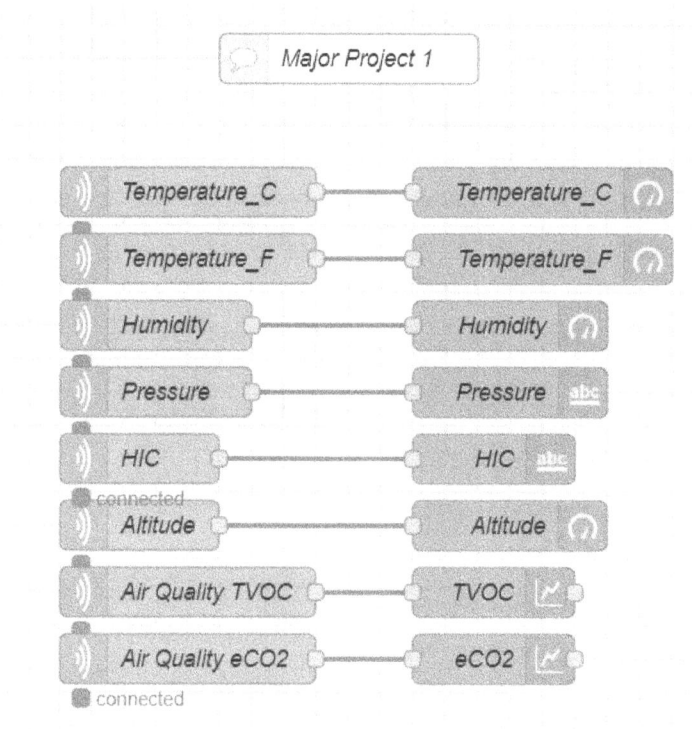

Figure 5.14 – The final Major Project 1 flow

The flow and dashboard setup is complete. Now, the only thing required is to deploy this flow. To do that, click the **Deploy** button in the top-right corner of the screen. If everything has been carried out as instructed, you will see that all your `mqtt` nodes have connected to the broker (indicated by the green dots).

Now, let us see how our project dashboard looks. For this, please power up the weather station by connecting the NodeMCU to a power source. Then open the Node-RED dashboard by typing the following IP address in your browser tab:

```
<--Raspberry Pi's IP Address-->:1880/ui
```

If you have multiple tabs, the first tab will be opened by default. Just click the menu icon at the top left and switch to the **IoT Weather Station** tab. If all the setup has been followed according to the mentioned steps, your dashboard will look similar to that shown in *Figure 5.15*.

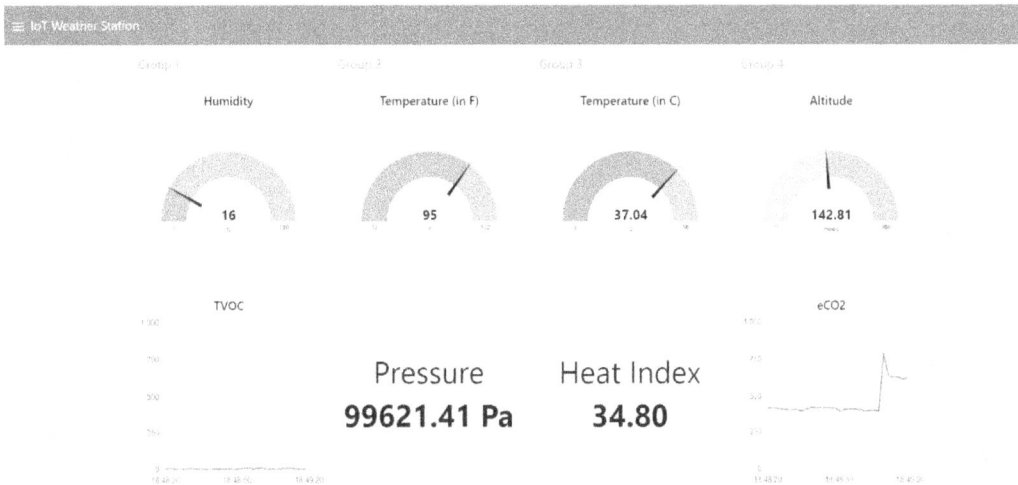

Figure 5.15 – The Major Project 1 Dashboard!

This marks the end of this section. In the next section, we will add some additional functionalities to our project. We will enable a simplified alert mechanism, which will send an email alert to a user when a sensor reading crosses a certain threshold value.

Additional functionality – email alerts

Now that we have the dashboard ready, we will add an additional feature to our project. We will be creating a simple email alert mechanism using Node-RED, which sends the user an alert whenever a particular defined event is triggered.

For instance, if we have set up an event that is triggered when the temperature of a room increases to 35 degrees Celsius, the system will automatically send an email alert to any email address of our choice.

To add the email feature, we will be using the email package of Node-RED. So, let's get started. Just follow the steps given below to implement this feature:

1. Go to **Manage palette** within the options menu, which you can open by clicking on the top-right icon on the home page. Please have a look at *Figure 5.16* for reference.

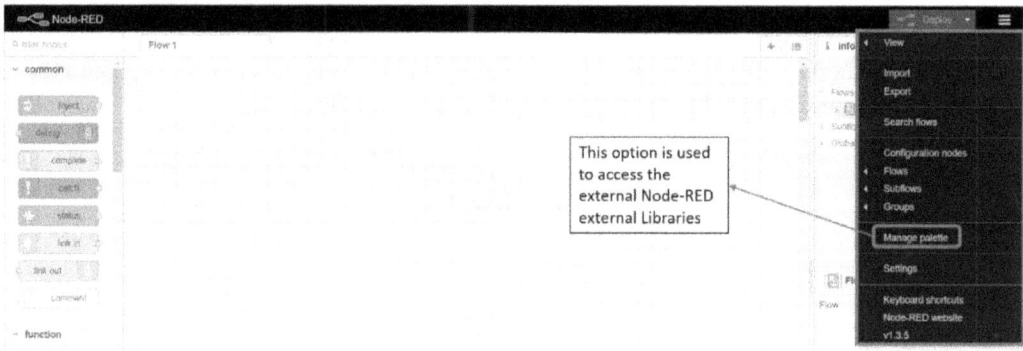

Figure 5.16 – Opening the Manage palette window in Node-RED

2. Search for the email package by typing `email` in the search box and install the `node-red-node-email` package (it should be at the top of the list of available packages). Please have a look at *Figure 5.17* for reference.

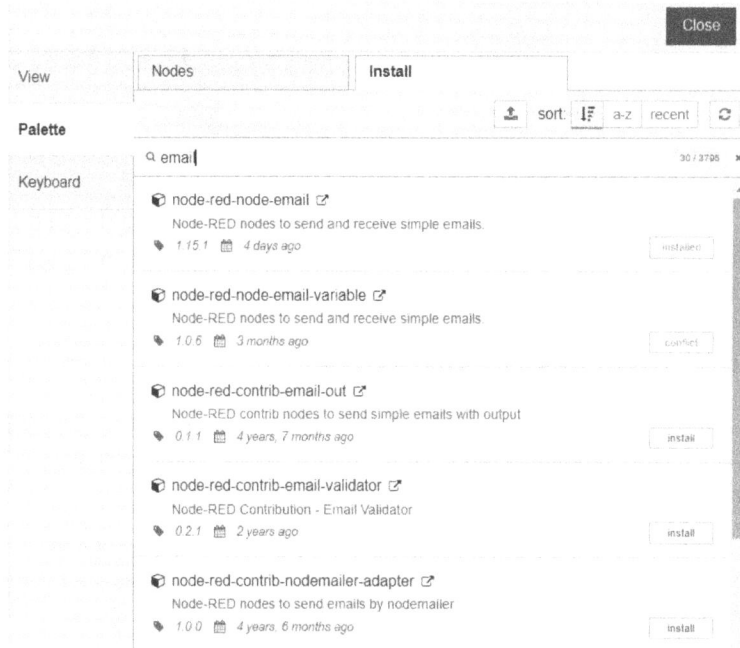

Figure 5.17 – Installing the Node-RED email package

3. Now, you can see the email-related blocks in the **social** section of the node manager. Please take a look at the following figure.

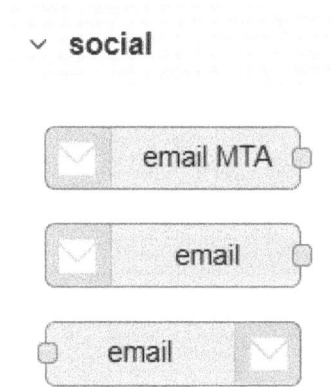

Figure 5.18 – The additional social section after installing the email package

4. Now, we are ready to create a new subflow that will enable us to send email alerts. Our alert system will work as follows: if the temperature goes above a threshold value, an email alert is generated saying that the alarm was triggered, and after it falls within the acceptable range again, another email saying that the alarm has been turned off is sent to the given email ID.

For this functionality, we require the following nodes:

- Two function nodes

- One email node (the connection dot on the left side indicates an output node)

- One debug node for debugging purposes

Once you drag these nodes into the workspace, it should look something like the following figure.

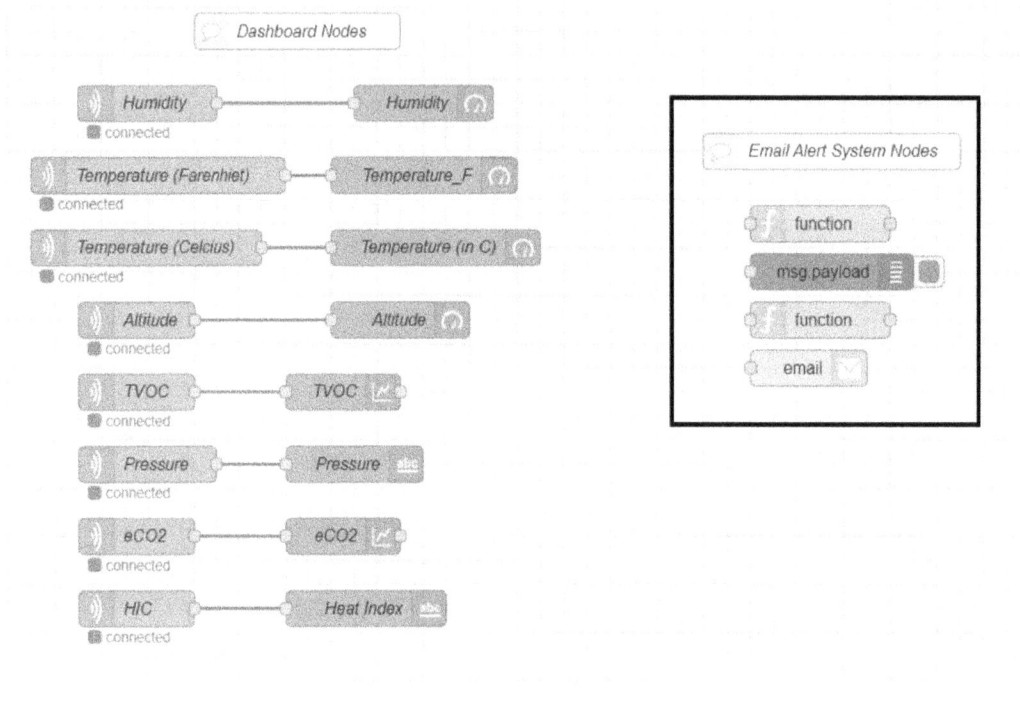

Figure 5.19 – The workspace with the nodes required to create the alert system

5. Next, we will need an email account that the email alerts will be sent to. Please note that one of the following conditions needs to be met if you are to use a Gmail account, with links to perform each of the given conditions:

    - Enable an application password (https://support.google.com/mail/answer/185833?hl=en).

    - Enable less secure access via your Google Account settings (https://support.google.com/accounts/answer/6010255?hl=en).

    Once this is done, we are ready to finally set up our flow.

6. We will be using two function nodes, one of which will be used to set up the logic and one of which will be used to create the actual body of the message. Please refer to the following code for each of the functions.

To alert the logic function, see the following:

```
var temperature=msg.payload;
var alert_flag=context.get("alert_flag");
if(typeof alert_flag=="undefined")
alert_flag=false;

if (temperature>35 && !alert_flag)
{
    alert_flag=true;
    msg.alert=1;
    context.set("alert_flag",alert_flag);
    return msg;
}
if (temperature<=35 && alert_flag)
{
    alert_flag=false;
    msg.alert=0;
    context.set("alert_flag",alert_flag);
    return msg;
}
```

The logic is fairly simple. We use the variable as an alert flag for the logic. If the temperature goes above 35 degrees Celsius, the alert flag is set, which in turn will be used to compose the alert message. Once the temperature falls below 35 degrees Celsius, the flag is set to `false` again, and another email alert will be sent saying that the temperature is back to normal.

Next, refer to the function that will create a simple alert message.

To create an alert message function, see the following:

```
var temp=msg.payload;
msg.to="<reciever email address>";

var d =new Date();
var message="";
if(msg.alarm)
{
    msg.topic="High Temperature Alert!";
    message="    ";
```

```
}
else
{
    message=" Temperature is back to normal now. The
current temperature is ";
    msg.topic="Temperature Alarm Reset.";
}
msg.payload="time:"+d+message +msg.payload;
return msg;
```

As depicted here, this will create a simple alert message for both the use cases discussed. The rest of the code is self-explanatory.

7. Next, we will be setting up the email node. A valid email address is required (if you are using Gmail, please carry out the aforementioned fixes to your settings) with the necessary credentials. Please refer to the following figure for the setup process.

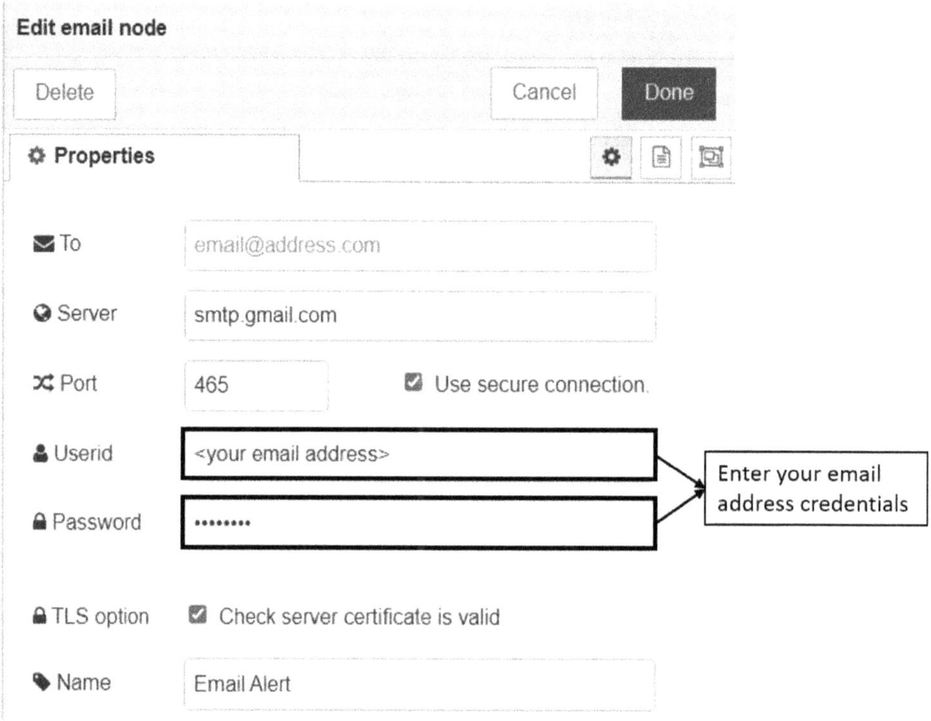

Figure 5.20 – Set up the email node

8. Once all the nodes are set up, connect the nodes that are shown in *Figure 5.21* to complete the project flow.

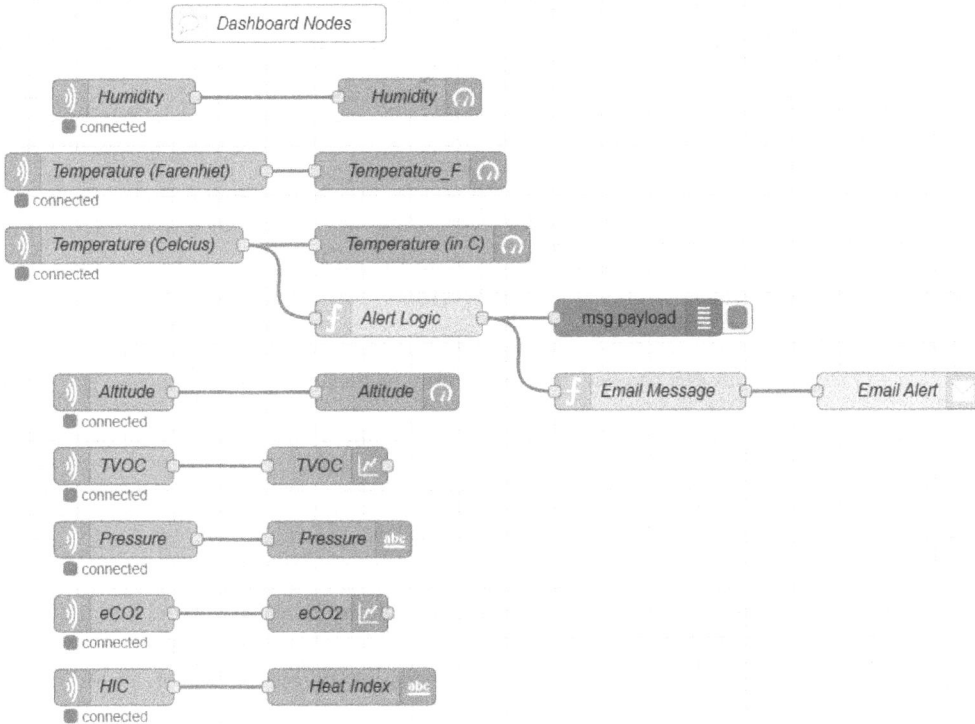

Figure 5.21 – The final project flow with an alert system

9. This completes the overall setup process.

Now, simply deploy the flow, and if you have followed this tutorial carefully, you should be able to send alert messages through Node-RED. A sample alert message looks something like the one that the following figure shows.

Figure 5.22 – A sample email alert message

This marks the end of this section and this chapter. You have successfully built your very first full-scale project: *a fully functional IoT weather station!* This project has dealt with getting you familiar with the monitoring capabilities of the Raspberry Pi and MQTT platforms.

There are several enhancements or improvements that can be made to this project:

- The first major enhancement would be additional nodes. The Pi and MQTT communication protocol can easily handle several devices simultaneously, so we can create multiple dashboards for different nodes.

- Next, we can add a data storage feature. The easiest way to achieve this is by configuring a MySQL database with Node-RED. Once this is done, we can easily store our sensor data in a database. This data can be used in the future to perform several different analysis tasks.

- We can even create forecasting models with a lightweight framework such as TensorFlow Lite, which will let us use our existing data to predict future values.

These are only a few of the many improvements that can be made to this project. The possibilities are endless.

Summary

This chapter guided you through building your very first major project using MQTT and Raspberry Pi. We built an IoT weather station based on the NodeMCU development board. We started by discussing the hardware requirements for the weather station and briefly introduced the sensors we would be using. Then, we moved on to interfacing these sensors to the NodeMCU board and proceeded with writing the code for our weather station. In the next section, we set up the Raspberry Pi for this project. We created our project dashboard on Node-RED and added a new alert feature to our project as well.

As discussed above, this is a monitoring-based project, which only allows us to fetch the data from a node and display it on the dashboard. In the next chapter, we will explore the control capabilities of this system by creating a dashboard to control a smart home device (relay) based on another very popular development board, the **ESP32 board**. As a bonus, we will even create a **printed circuit board (PCB)** for our project to give it a more professional and finished look.

6

Major Project 2: Smart Home Control Relay System

In this chapter, we will be creating yet another full-scale project using the concepts we learned in the initial chapters of this book. The main objective of this chapter is to create a smart home device for controlling wall switches using the Node-RED dashboard hosted on the Raspberry Pi. The device will be based on the popular ESP32 development board.

This chapter is a step-by-step tutorial to build this project from scratch. We will be covering the following aspects of the project:

- Hardware requirements and setup
- Code explanation
- Raspberry Pi setup
- Project enhancements

For this project, we will be preparing a PCB instead of creating the circuit on a breadboard for a more finished and professional look. The final hardware is shown in *Figure 6.1*.

Figure 6.1 – Smart home control relay system (ESP32)

In this figure, we can see that the hardware has four relays, and the functionality or features include the following:

- Control via MQTT or any other mode supported by the ESP32
- Manual control functionality (control using conventional switches)
- An alarm using an on-chip buzzer

Please note that all the code resources will be available in the GitHub repository for this book. The link for the repository is in the *Preface* section.

Important Note

Please note that basic soldering skills are a prerequisite for this project. You will need to solder the components on the PCB. All the components are through-hole, so a beginner-level solder iron will also be sufficient for this task.

The first thing we will do is go through the hardware requirements and setup process. We will briefly learn about all the components used in this project, followed by the PCB design and assembly.

Hardware requirements and setup

This project has been chosen to demonstrate the control capabilities of our Raspberry Pi MQTT system. Hence, there will be no sensors used in this project. But that is very possible and extremely easy to implement. In fact, that can be a challenge for you: build a project with both monitoring and control components.

For now, the components required to build this project are as follows (*Figure 6.2*):

- ESP32 development board
- 5V non-latching relay
- Hi-Link 5V power supply
- Two resistors – 10k ohm and 330 ohm
- Two-pin terminal connectors
- BC547 transistor
- 1N4007 diodes
- Buzzer
- LEDs

Figure 6.2 – Required components to build this project

Now, we will briefly go through each major component's role, as we did for the first project.

ESP32 development board

We have chosen the ESP32 development board as the brains for this project (*Figure 6.3*). One major reason for that is that it gives us the option to use Bluetooth, which is a universal communication protocol (especially in smartphones), to control our IoT devices. Even though we are not going to use Bluetooth in this project, it is possible to switch this project to a Bluetooth-based home automation system.

Figure 6.3 – ESP32 development board

Moreover, it is more powerful than the ESP8266 chip on the Node MCU and it has much more pin configuration possibilities. The salient features of this device are as follows:

- It has 520 KB of SRAM, making it suitable for numerous applications.

- It boasts a hybrid chip to support both Wi-Fi and Bluetooth wireless communication protocols.

- It supports numerous power management modes, making it perfect for low-power applications.

- It has a 4 MB flash memory.

- It possesses an onboard antenna on the microcontroller PCB for long-range wireless communications.

5V non-latching relay

We want to control home switches, for which we will be requiring a switch that can be controlled through our development board. However, there is no provision to connect a switch to the microcontroller directly (it would fry the chip!). So, what is the solution? Relays!

Relays are the devices that act as these switches. Their primary task is to close and open an AC connection when it is triggered by a low-power DC signal, which a microcontroller is capable of doing. We will be adding relays in addition to our conventional home switches so that we can control them using MQTT.

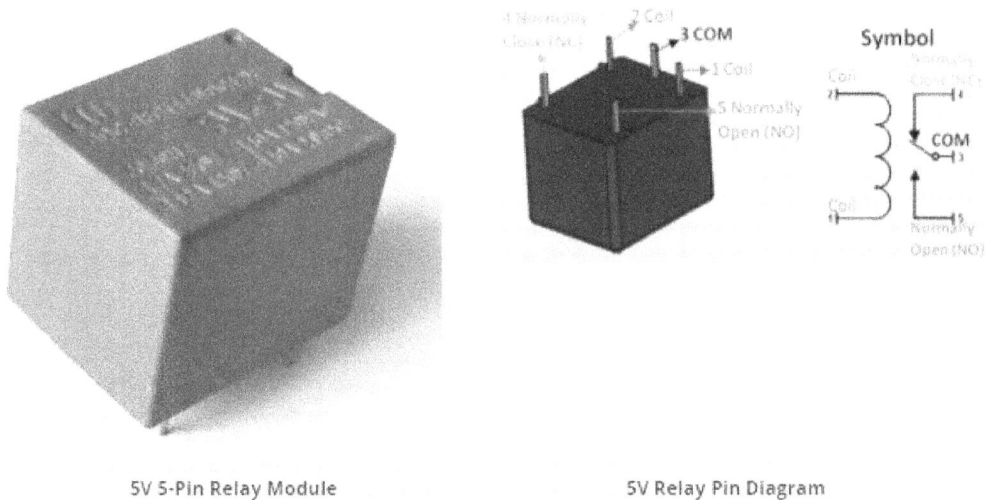

5V 5-Pin Relay Module   5V Relay Pin Diagram

Figure 6.4 – Relay module with pin configuration

Primarily, when dealing with mains power, **solid state relays** (**SSR**) are preferred, as they have a number of advantages over conventional non-latching relays. But as this is a hobby project, we will be using 5V non-latching relays for this project. Please note that SSRs are a lot more expensive than normal relay modules. The features of the relay that we will be using are as follows:

Trigger Voltage	5V DC
Trigger Current	70mA
Maximum AC load current	10A @ 250/125V AC
Maximum DC load current	10A @ 30/28V DC
Operating time	10msec
Release time	5msec
Maximum switching	300 operating/minute

Figure 6.5 – Relay module (5V-10A) specifications

5V Hi-Link power supply

The Hi-Link 5V switch power supply is a very popular and low-cost step-down power supply module. The primary goal of this component is to convert the conventional 120-230V AC supply to a 5V DC output. It is available in different power and voltage ratings, but we choose the simplest of them, which has a power rating of 5 W (Power = Voltage * Current), so it supports a maximum current of 1 Ampere.

The package we are using is developed specifically for use on different printed circuit boards, making it the perfect choice for custom DIY circuits. Being a switching source, it handles all the voltage fluctuations in the voltage grid internally. It is widely used in home automation and smart home applications. The component will look something like this:

Figure 6.6 – Hi-Link power supply (5V – 5 W)

Salient features of this component are as follows:

- Available in a thin and small package.
- Supports an input voltage in the 90–264V range.
- Provides a steady output with low ripple and noise.
- It has inbuilt safety features, such as short circuit protection and output overload.
- Low power consumption, environmental protection, and no-load loss < 0.1 W.

Miscellaneous components

Apart from the components listed previously, several passive components are required for this project. These components will be discussed in this subsection.

First, the BC547 transistors (4) and resistors (10k-ohms) are required to complete the relay circuits. In order to operate a relay through our development board, we require these components. The transistor acts as a switching circuit and helps change the relay state from our development board's digital pin.

The circuit diagram is shown in *Figure 6.7*:

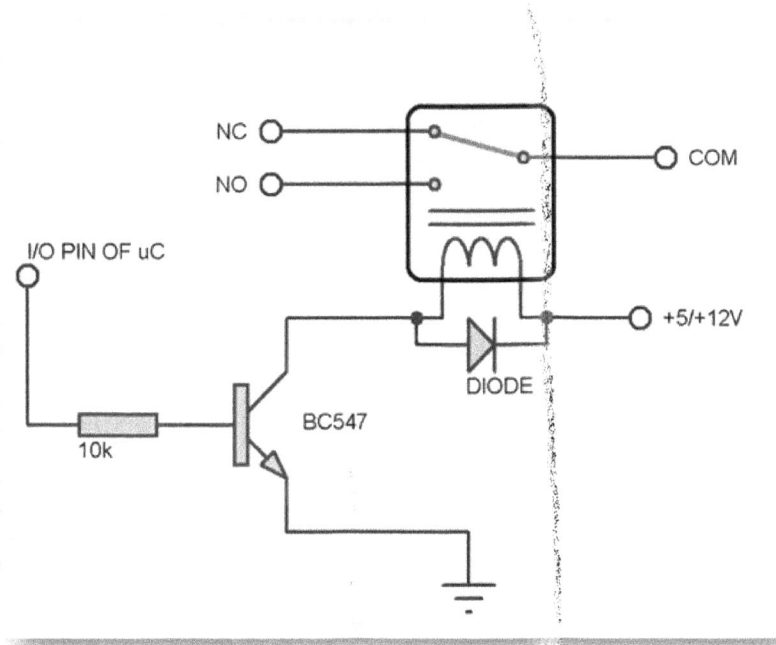

Figure 6.7 – Relay module connection diagram

Moreover, the LEDs are to show the status for different services (Wi-Fi connection, MQTT server connection, and so on) of our ESP32. The 330-ohm resistors are used to connect the LEDs to the ESP32 digital pins.

The buzzer can be used as an alarm when something goes wrong. This concludes the explanation of all the hardware components we are going to use in this project. Next, we will cover how to assemble everything.

Hardware setup (PCB design and circuit)

As mentioned in the introduction, we will be developing a PCB for this project. But PCB designing requires a lot of skill, experience, and knowledge, which is a separate discussion topic itself. So, we will be using an already available *open source* design that was developed by a tech YouTuber called *Sachin Soni* (the channel is called *techiesms*).

Please refer to *Figure 6.8* for the PCB image. Note that the design is readily available on the popular PCB development website *EasyEDA*.

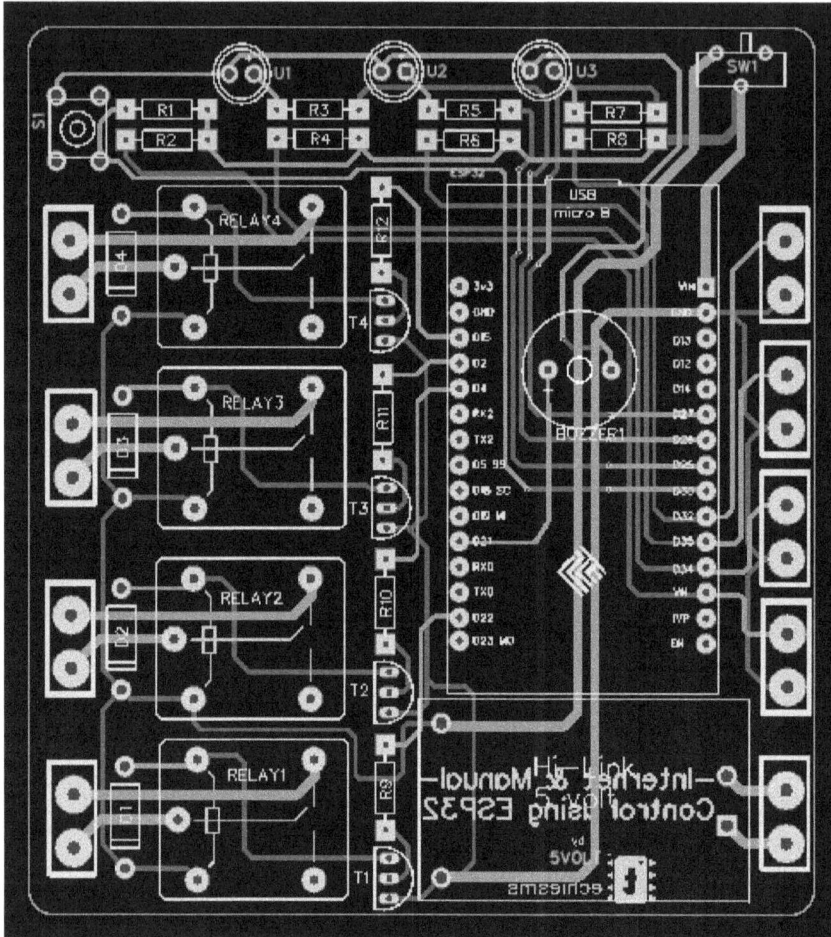

Figure 6.8 – Project PCB design schematic

You can just download the PCB Gerber file and order your own set of PCBs from any of the popular PCB providers (such as JLC PCB). It will take some time to get your hands on those PCB boards, and the amount of time depends on the country you live in. Once you have the PCB and all the components, we are ready to build our project.

For those who are new to soldering, I have added a link to a YouTube video that walks you through the basics of soldering in the reference links. Please go through that video if you have never soldered anything before. I suggest you buy some extra components to practice on a test board first. Once you get the hang of it, just solder all the components to the printed circuit board for this project.

If you do not have access to a soldering iron or do not know how to use one, you can create a simple circuit on a breadboard or a PCB using the schematic diagram shown in the following image:

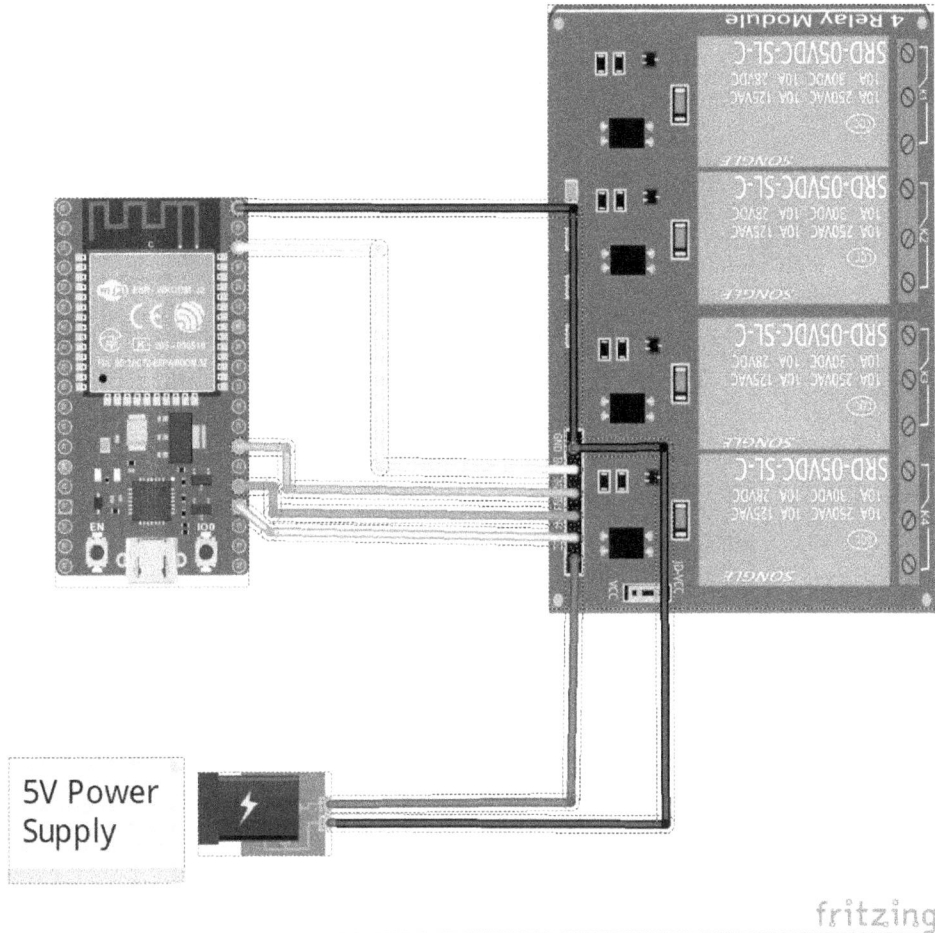

Figure 6.9 – Basic breadboard project schematic

Please note that you will need to supply the relay board externally with a 5V supply as the ESP32 has a maximum voltage output of 3.3V.

> **Important Note**
>
> Soldering irons can seriously injure you if not used correctly. If you are a minor, please do it under adult supervision.

You have already seen in the first section what the final PCB looks like (refer to *Figure 6.1*). Please refer to *Figure 6.10* to see how the circuit is connected:

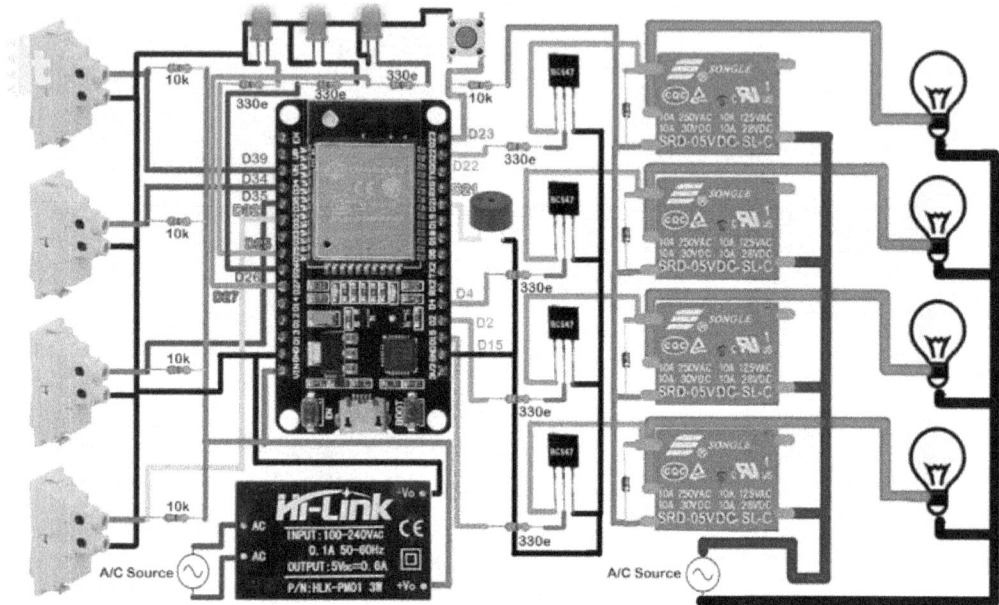

Figure 6.10 – Project connection schematic diagram

In the next section, we will be writing and uploading the project code to our ESP32 board.

Code explanation

The hardware setup is complete and the PCB is ready. It is time to write some code for our project. This code will be written in the Arduino IDE for our ESP32 development board.

The code will have the following tasks to perform:

1. Connect to the pre-configured Wi-Fi network.

2. Connect to the MQTT broker hosted on the same network (on the Pi, in our case).

3. Pin initialization for the GPIO pins we will be using in this project.

4. Subscribe to various switch topics.

5. Reconnect to the MQTT server if it disconnects.

6. Develop a logic to control the relays based on the payload received on those topics (in the callback function).

Points 4, 5, and 6 will run indefinitely (part of the `loop()` function). The code is available on the GitHub repository of this project. Now, we will divide the code into parts to make it easier to understand, as we did before.

To import the required libraries use this code:

```
// Importing the required Libraries
#include <WiFi.h> //WiFi functionality access for ESP32
#include <PubSubClient.h>    // Enables the use of MQTT
```

We will first import the required libraries, and in our case there are only two. The first one is the Wi-Fi library, which enables our ESP32 to connect to any Wi-Fi network. The second one is the Pubsubclient library, which gives ESP32 the ability to connect to MQTT brokers.

Please note that there are no sensors used in this project, unlike in the first major project, so no additional libraries are required. Hence, we can run both projects on the same board together (this will be given as an assignment later).

Next, we will initialize the necessary variables and objects.

To initialize variables and object, use this code:

```
// WiFi and MQTT Credentials
const char* ssid = "wifi_ssid";
const char* password = "wifi_password";
const char* mqtt_server = "Pi's ip address";

// Other Variable and object declarations
int relay1 = 15;
int relay2 = 2;
int relay3 = 4;
int relay4 = 22;

WiFiClient espClient;
PubSubClient client(espClient);
```

Next, we have to define the constants, variables, and objects for the project. That includes the *Wi-Fi and MQTT credentials* and *Wi-Fi and MQTT client initialization*.

Moreover, we will assign the pins to which the relays are connected as variables as they will be used multiple times in the code.

For the Wi-Fi setup function, use this code:

```
// Custom function for Wifi connection establishment
void setup_wifi()
{
  delay(10);
  Serial.println();
  Serial.print("Connecting to ");
  Serial.println(ssid);
  WiFi.mode(WIFI_STA);
  WiFi.begin(ssid, password);
  while (WiFi.status() != WL_CONNECTED)
  {
    delay(500);
    Serial.print(".");
  }
  randomSeed(micros());
  Serial.println("");
  Serial.println("WiFi connected");
  Serial.println("IP address: ");
  Serial.println(WiFi.localIP());
}
```

The `setup_wifi` function is used to connect to the Wi-Fi network, the credentials of which we provided as constant variables, `ssid` and `password`.

In the MQTT callback function, add this to read the received messages:

```
void callback(char* topic, byte* message, unsigned int length)
{
  // Reading the received messages
  Serial.print("Message arrived [");
  Serial.print(topic);
  Serial.print(". Message: ");
  String messageTemp;

  for (int i = 0; i < length; i++) {
    Serial.print((char)message[i]);
```

```
    messageTemp += (char)message[i];
  }
  Serial.println();
```

In this section, we first check for any messages that are received on a topic and store it in a variable. The next sections are the core logic for this project, checking messages on particular topics, and setting the relays accordingly.

The below code snippet describes the code logic to control Switch 1 through our dashboard.

```
  // Relay Control Logic starts here.

  if (String(topic) == "IoTSmartSwitches/Switch1") {
    Serial.print("Changing output to ");
    if(messageTemp == "1"){
      Serial.println("Switch 1 turned On");
      digitalWrite(relay1, HIGH);
    }
    else if(messageTemp == "0"){
      Serial.println("Switch 1 turned off");
      digitalWrite(relay1, LOW);
    }
  }
```

This is the main code block, which repeats for every single relay module pin. First, we check for the topic, and if it matches, we check the message we have received. If we get **1**, we turn on the relay and if we receive **0**, we turn it off. Please note that if any other message is received on this topic, there will be no change in the relay state.

Similarly, the code snippets below describe the code logic to control switches 2-4 through our dashboard.

```
  else if (String(topic) == "IoTSmartSwitches/Switch2") {
    Serial.print("Changing output to ");
    if(messageTemp == "1"){
      Serial.println("Switch 2 turned On");
      digitalWrite(relay2, HIGH);
    }
    else if(messageTemp == "0"){
      Serial.println("Switch 2 turned off");
```

```
        digitalWrite(relay2, LOW);
    }
}
```

The code block for Relay 2 is the same, with only the change in the topic name and the relay pin that is controlled.

```
    else if (String(topic) == "IoTSmartSwitches/Switch3") {
      Serial.print("Changing output to ");
      if(messageTemp == "1"){
        Serial.println("Switch 3 turned On");
        digitalWrite(relay3, HIGH);
      }
      else if(messageTemp == "0"){
        Serial.println("Switch 3 turned off");
        digitalWrite(relay3, LOW);
      }
    }
```

The code block for Relay 3 is the same, with only the change in the topic name and the relay pin that is controlled.

```
    else if (String(topic) == "IoTSmartSwitches/Switch4") {
      Serial.print("Changing output to ");
      if(messageTemp == "1"){
        Serial.println("Switch 4 turned On");
        digitalWrite(relay4, HIGH);
      }
      else if(messageTemp == "0"){
        Serial.println("Switch 4 turned off");
        digitalWrite(relay4, LOW);
      }
    }
  }
```

With the logic defined for all four relays, we will get real-time feedback for our switches. We can control their state just by sending out messages to their respective topics.

> **Important Note**
>
> Please keep in mind that this code assumes that your board pins are active high. But some ESP32 board models (mine, for one) have active low pins, so you would have to exchange the `digitalWrite` commands for each case.

Here's the MQTT reconnect function:

```
void reconnect()
{
  // Loop until we're reconnected
  while (!client.connected()) {
    Serial.print("Attempting MQTT connection...");
    // Create a random client ID
    String clientId = "ESPClient-";
    clientId += String(random(0xffff), HEX);
    // Attempt to connect
    if (client.connect(clientId.c_str()))
    {
      Serial.println("connected");
      client.publish("outTopic", "Reconnected!");

      // Subscribe to all the relevant topics
      client.subscribe("IoTSmartSwitches/Switch1");
      client.subscribe("IoTSmartSwitches/Switch2");
      client.subscribe("IoTSmartSwitches/Switch3");
      client.subscribe("IoTSmartSwitches/Switch4");

    }
    else
    {
      Serial.print("failed, rc=");
      Serial.print(client.state());
      Serial.println(" try again in 5 seconds");
      // Wait 5 seconds before retrying
      delay(5000);
    }
  }
}
```

The *MQTT reconnect function* is in place to establish a connection to the broker again in case there are some issues with the ESP32 development board. These issues could be anything from internet connectivity issues to hardware failure.

This function tries to reconnect to the broker every 5 seconds and once the connection has been re-established, it will *resubscribe to all the individual switch topics*.

Here's the `setup` function:

```
void setup()
{
  // put your setup code here, to run once
  Serial.begin(115200);

  setup_wifi();
  client.setServer(mqtt_server, 1883);
  client.setCallback(callback);

  client.subscribe("IoTSmartSwitches/Switch1");
  client.subscribe("IoTSmartSwitches/Switch2");
  client.subscribe("IoTSmartSwitches/Switch3");
  client.subscribe("IoTSmartSwitches/Switch4");

  pinMode(relay1, OUTPUT);
  pinMode(relay2, OUTPUT);
  pinMode(relay3, OUTPUT);
  pinMode(relay4, OUTPUT);

  digitalWrite(relay1, LOW);
  digitalWrite(relay2, LOW);
  digitalWrite(relay3, LOW);
  digitalWrite(relay4, LOW);


}
```

The setup function does the following:

- Opens a serial connection port with a baud rate of 115200

- Connects to Wi-Fi and establishes a connection to our Pi (MQTT broker)

- Subscribes to all four switch MQTT topics
- Sets the relay pins' mode to *output* and set all those pins to *Low*

Here's the `loop` function:

```
void loop() {
  // put your main code here, to run repeatedly
  if (!client.connected()) {
    reconnect();
  }
  client.loop();
}
```

The `loop` function has only one task: to check if the MQTT connection is intact, and if it is not, then run the reconnect function.

This marks the end of the code explanation section. In the next section, we will set up our Raspberry Pi for this project. That includes the following tasks:

- Setting up the MQTT broker (already done)
- Creating a Node-RED flow
- Creating a dashboard for this project

Looks like we have our work cut out for us. Let's get started!

Raspberry Pi setup

The Raspberry Pi will be the host for the local MQTT broker in this project and also the dashboard hosting device. The dashboard for this project will be created using Node-RED and the Node-RED dashboard module, which will both be running on the Raspberry Pi.

Setting up MQTT and Node-RED on the Raspberry Pi has already been covered, so we will start with the Node-RED setup portion straight away. The first step is to start Node-RED after booting up your Pi.

Just open a new terminal on your Pi and type in the following command:

```
node-red-start
```

Please refer to *Figure 6.11* for reference:

```
pi@raspberrypi:~ $ node-red-start

Start Node-RED

Once Node-RED has started, point a browser at http://192.168.1.22:1880
On Pi Node-RED works better with the Firefox or Chrome browser

Use   node-red-stop                           to stop Node-RED
Use   node-red-start                          to start Node-RED again
Use   node-red-log                            to view the recent log output
Use   sudo systemctl enable nodered.service   to autostart Node-RED at every boot
Use   sudo systemctl disable nodered.service  to disable autostart on boot


To find more nodes and example flows - go to http://flows.nodered.org


Starting as a systemd service.
6 Apr 15:36:26 - [info]
Welcome to Node-RED
===================
6 Apr 15:36:26 - [info] Node-RED version: v2.2.2
6 Apr 15:36:26 - [info] Node.js  version: v14.19.1
6 Apr 15:36:26 - [info] Linux 5.10.92-v7l+ arm LE
6 Apr 15:36:27 - [info] Loading palette nodes
6 Apr 15:36:29 - [info] Dashboard version 3.1.6 started at /ui
6 Apr 15:36:29 - [info] Settings file  : /home/pi/.node-red/settings.js
6 Apr 15:36:29 - [info] Context store  : 'default' [module=memory]
6 Apr 15:36:29 - [info] User directory : /home/pi/.node-red
6 Apr 15:36:29 - [warn] Projects disabled : editorTheme.projects.enabled=false
6 Apr 15:36:29 - [info] Flows file     : /home/pi/.node-red/flows.json
6 Apr 15:36:29 - [info] Server now running at http://127.0.0.1:1880/
6 Apr 15:36:29 - [warn]
-------------------------------------------------------------------
Your flow credentials file is encrypted using a system-generated key.
If the system-generated key is lost for any reason, your credentials
file will not be recoverable, you will have to delete it and re-enter
your credentials.
You should set your own key using the 'credentialSecret' option in
your settings file. Node-RED will then re-encrypt your credentials
file using your chosen key the next time you deploy a change.
```

Figure 6.11 – Starting Node-RED on the Pi

This command will start your Node-RED and give you the IP address from which you can access the Node-RED editor from any device connected to the same network as the Pi. Once you open Node-RED, the next step is creating the flow and, in turn, the dashboard for this project. As this is a control project, the dashboard will be fairly simple, consisting of just four switches. There are some things that can be added to improve this project, but we will stick to the basics to make things easier for beginners.

Just follow the step-by-step instructions to set up the entire Node-RED environment for this project (including both the flow creation and dashboard setup):

1. On the Node-RED home screen, just create a new flow following the instructions provided in *Figure 6.12* to get a new and blank workspace.

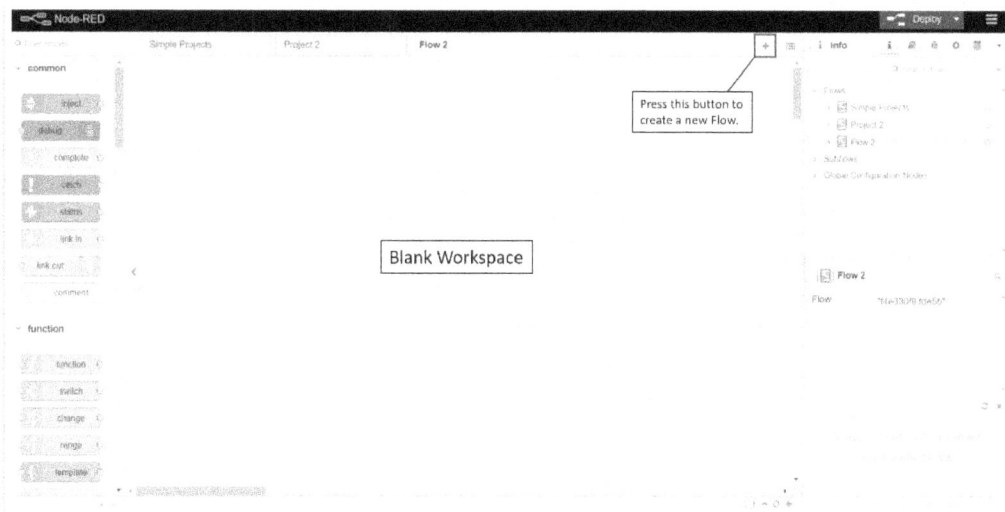

Figure 6.12 – Creating a new flow in Node-RED

2. Next, we need to install a new extension for better dashboard switches. So, just go to additional options and click **Manage Pallete**, and then go to the **install** tab on the new window that opens. Search for `node-red-contrib-ui-multistate-switch` and download the first extension.

3. Create a new flow; you can give it the name of your choice. Once that is done, just drag out the following nodes onto the workspace (as shown in *Figure 6.13*):

    • Four `mqtt out` nodes

- Four `multistate switch` nodes

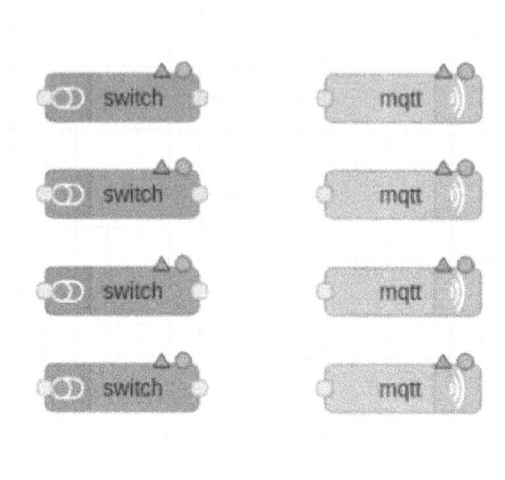

Figure 6.13 – Nodes used for this project

4.   Please note that we will create a simple dashboard with four switches for this project. We will use two-state switches, each with on and off buttons. The dashboard layout is simple and shown in *Figure 6.14*.

After the layout setup is done, the next step is to set up each of the nodes.

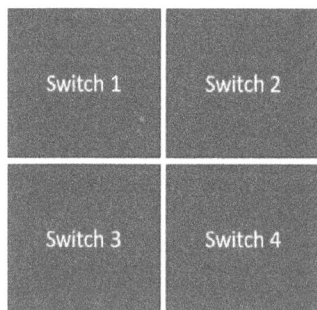

Figure 6.14 – Node-RED dashboard layout setup

5. The setup instructions for both the `mqtt out` and `switch` nodes are shown in *Figure 6.15*.

    The following information is required to be filled for the `mqtt out` node:

    - Select the MQTT broker

    - Topic name according to the switches

    - Name of the node (name appearing on the workspace)

    The following information needs to be entered for the `switch` node:

    - Name of the node

    - **Group** for the dashboard layout setup

    - **Label** string (the string that appears on the dashboard)

    - Set **Appearance** to rounded (based on your preference)

    - Add the labels as shown in *Figure 6.15*:

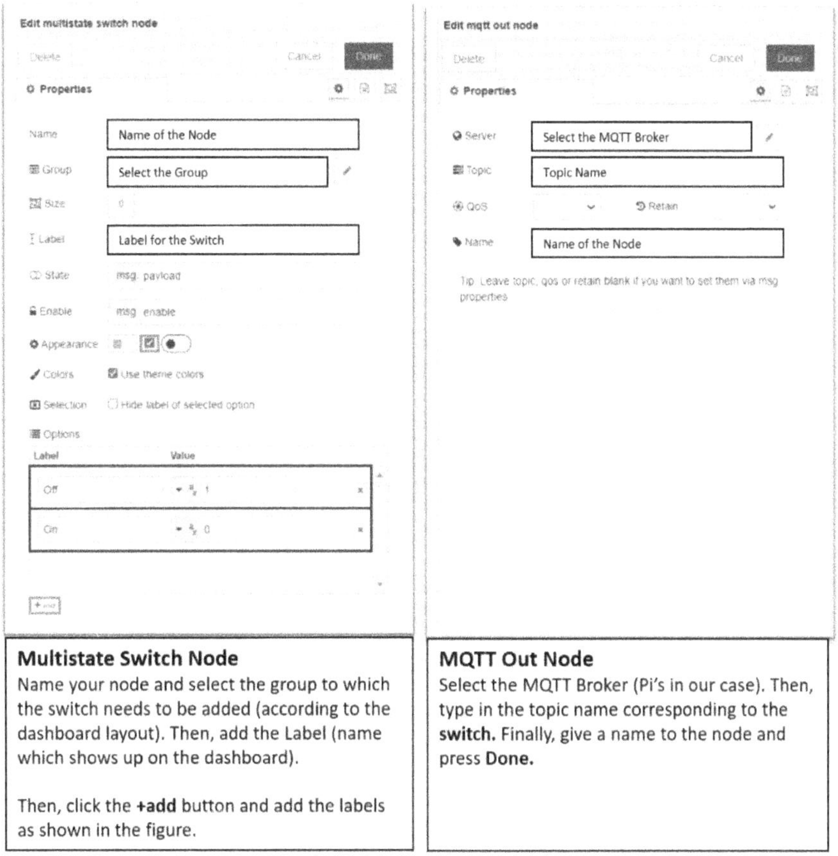

Multistate Switch Node
Name your node and select the group to which
the switch needs to be added (according to the
dashboard layout). Then, add the Label (name
which shows up on the dashboard).

Then, click the **+add** button and add the labels
as shown in the figure.

MQTT Out Node
Select the MQTT Broker (Pi's in our case). Then,
type in the topic name corresponding to the
switch. Finally, give a name to the node and
press **Done.**

Figure 6.15 – Node setup for this project

6. Now that the nodes have been set up for our project, we will just connect the nodes according to *Figure 6.16*:

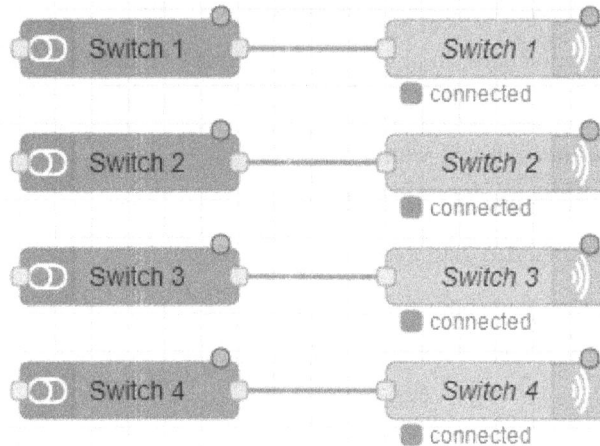

Figure 6.16 – Final Project 2 flow

Once the flow has been set up, all that we need to do is deploy the node. To do that, just click the **Deploy** button in the top-right corner of the screen. Once that is done, you should see that the blue dots on each of our project's nodes have vanished.

If everything has been done as instructed, you will see that all your MQTT nodes have connected to the broker (indicated by the green dot). Hence, we have successfully completed the Node-RED setup. Now all that is left for us to do is test the dashboard.

7. Now, let's see how our project dashboard looks. For this, please power on the PCB by connecting it to an AC power supply (powering the ESP32 won't work).

Then open the Node-RED dashboard by typing the following IP address on your browser tab:

```
<Raspberry Pi's IP Address>:1880/ui
```

If you have multiple tabs, the first tab will be opened by default. Just press the menu icon in the top left and switch to the **IoT Smart Switches** tab. If you have followed all the setup steps, your dashboard will look similar to the one shown in *Figure 6.17*:

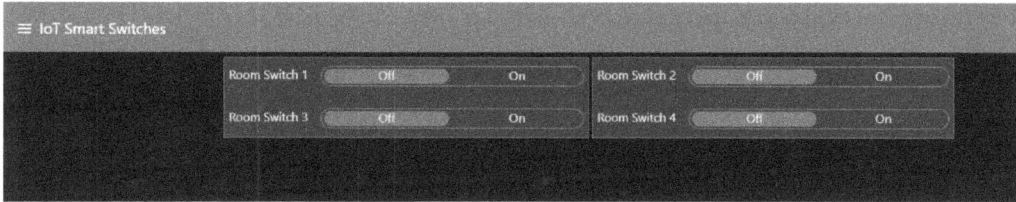

Figure 6.17 – Project 2 dashboard (in dark mode!)

As you can see, you have control of all the four relays here. Now, all that's required is to test this system. Just turn on any switch and you should see an instantaneous change in the state of one of the relays. There is a very high speed and low latency connection established between the Pi and the ESP32 project.

You should be able to control all the devices connected to those relays through this dashboard.

Congratulations! You just completed the second and the final major project for this book. Now, you are aware of the control capabilities of this system as well. But there is always room for improvement. Several improvements can be made as far as this project is concerned. In the next section, we will be discussing what exactly can be done to make this project better than it already is.

Project enhancements

Project enhancement is a crucial part of project development. We always strive to make things better than they already are. This case is no different.

There are several possible enhancements, both on the hardware and software. Let's walk through some of them:

1. The first and main hardware enhancement is adding manual feedback to our system. In fact, the PCB supports it.

    The current system, as it stands, does not allow the user to use manual switches, and even if we managed to use them, we cannot get their statuses (that is, at any given moment, we can't see the state of a switch). But the PCB we are using has a special function: it can provide the current state of any connected application (on or off) on particular ESP32 digital pins. Hence, we can get feedback. Please refer to the circuit diagram for the PCB to see how you need to connect the switch wires to the PCB.

The portion on the PCB that helps achieve this is marked in *Figure 6.18*:

Figure 6.18 – Screw terminals that add manual feedback

Now, as far as the code part is concerned, here is a code snippet that will give you some idea about how to implement this in your existing project. This is an assignment for you: *implement manual control into your IoT smart home system.*

Manual Automation Code Snippet

```
#define S1 32
#define S2 35
#define S3 34
#define S4 39

// You can access the LEDs and Buzzer through this pins.
#define LED1 26
#define LED2 25
#define LED3 27
#define Buzzer 21
```

```
void Call_ManualAutomation()
{
  Serial.println("Manual Automation");
  digitalWrite(R1, digitalRead(S1));
  Serial.println("Relay-1: ");
  Serial.println(digitalRead(S1));
  delay(1);
  digitalWrite(R2, digitalRead(S2));
  Serial.println("Relay-2: ");
  Serial.println(digitalRead(S2));
  delay(1);
  digitalWrite(R3, digitalRead(S3));
  Serial.println("Relay-3: ");
  Serial.println(digitalRead(S3));
  delay(1);
  digitalWrite(R4, digitalRead(S4));
  Serial.println("Relay-4: ");
  Serial.println(digitalRead(S4));
  delay(1);
}
```

2. The next possible improvement that can be made is the development of a mobile application to control the switches. Technically, we can still open the Node-RED dashboard on our phone browser and control the relays from there, but a mobile application is a more finished and better solution.

    A number of no-code app development platforms are available. The best and the easiest to use is the MIT App Inventor. You can find the link for this in the references section. It allows you to develop apps using drag-and-drop components and a no-code setup process. I have also included a link to a tutorial video on how to use this platform in the references section.

    Here is an *app development exercise* for you.

    Develop an application to control the relays on our project's PCB using MIT App Inventor. The following links will help you get started:

    *   MIT App Inventor: https://appinventor.mit.edu/

    *   Tutorial video: https://highvoltages.co/iot-internet-of-things/how-to-mqtt/how-to-make-mqtt-android-application-using-mit-app-inventor/

3.   Another possible enhancement in this project would be the addition of a global MQTT broker. Currently, the project can only be used within the local network (your Wi-Fi connection).

In order to be able to control these switches from anywhere in the world, we would require an MQTT broker that is hosted online and accessible on any network. This opens a whole new world of possibilities: the addition of multiple devices, monitoring components, and so on. We can achieve this in a number of ways. These will be discussed in detail in the next chapter.

Summary

We developed our second full-scale project in this chapter, creating a smart relay system based on ESP32 and operated over MQTT. We will walk through the key points we covered in this chapter just to refresh your memory.

We started with the hardware requirements and setup of those components. This includes setting up our system and connecting all the components on a custom PCB. Next, we moved on to the explanation of the code, wherein we broke the code into several snippets to make it easier to understand. After that, we set up the project dashboard on the Raspberry Pi. Finally, we moved onto the demonstration part of the project to see our project in action.

In the next chapter, we will cover how to take this concept even further by taking the MQTT broker global so that we can access our devices through MQTT from anywhere without the constraint of local network coverage.

Part 3: How to Take Things Further – What Next?

The two projects we've covered gave you a hands-on experience in developing innovative end-to-end IoT projects. But how do we scale these projects and move to the next level? You will find out in this part.

This part comprises of the following chapters:

7

Taking Your MQTT Broker Global

The previous two chapters covered two fully functional prototype projects so that you can get hands-on experience on how to build IoT projects. So, what next?

Now that you have seen the potential of IoT and the hardware we used – the Raspberry Pi – you are ready to learn how to utilize these technologies best. In this chapter, we will talk about the MQTT broker that we have currently hosted on the Raspberry Pi. This gives us access to it within a local network only. But what if we can access it over the internet?

Figure 7.1 – MQTT meets the internet!

In this chapter, we will cover the following topics:

- Establishing the advantages of a global MQTT broker
- How to take your broker global

Everything discussed in this chapter is not always required, so if you just want to create small projects for your home, I suggest sticking with the Raspberry Pi instead of using the options presented in the coming sections.

> **Important Note**
>
> Please note that most of the options that we will be discussing in this chapter will be paid for (monthly subscriptions mostly). This is optional and you can avoid this chapter if you are just a beginner and do not intend to scale your project beyond your own home.

Establishing the advantages of a global MQTT broker

There are several advantages of using an MQTT broker that is hosted on a server or machine with internet access. We will look at some of these in this section:

- **Accessibility**: You can access the devices connected to your broker from anywhere and with any device, so long as you have an active internet connection.

- **Scalability**: When you use a Pi, you have limited coverage. However, that is not the case with an online MQTT broker. You can easily connect multiple smart devices, which are present in different locations.

- **Efficient channel use**: This method supports multiple devices, allowing you to proficiently utilize channels.

- **Large-scale compatibility**: If you want to use this at a larger scale (for example, you have more than 10,000 devices in your ecosystem), you can easily choose a new plan, which gives you access to more storage and more topics (larger channel bandwidth).

- **Cost efficient**: We can keep using the local brokers (hosted on low-cost devices such as the Raspberry Pi) so that we can still get a control interface for our project. Also, we get certain functionalities such as storage capabilities without increasing our costs by paying for extra storage space on online servers.

These benefits are just some of the reasons why we should switch to an online broker. Moreover, we already have the advantages that MQTT has to offer over other communication protocols, such as HTTP.

Now, let's discuss the options we have at present to switch to an online MQTT broker and how choosing a particular option will impact your existing devices.

How to take your broker global

In this section, we will discuss the two major options that we have to grant internet access to our existing projects.

There are two ways to obtain access to a global MQTT broker:

- **Online MQTT brokers**: Several online MQTT brokers provide you with a ready-to-use MQTT broker (credentials are provided). These are available on a subscription basis (monthly, quarterly, or annually).

  Some popular sites for this are HiveMQ, Paho, CloudMQTT, and Adafruit IO. These can be seen in the following diagram:

Figure 7.2 – Some popular online MQTT brokers

We will be using the HiveMQ platform to test this later. We will use the free plan provided by the platform, wherein you will be given the following features:

Capacity

MQTT Client Sessions:	100
Data Traffic:	10 GB
Data Retention Time:	3 Days
Max Message Size:	5 MB

UPGRADE CLUSTER

Figure 7.3 – Free plan provisions

- **Virtual Server**: If you want more customization and additional features, you can get a virtual server from AWS, Azure, Google Cloud Platform, or Digital Ocean.

  After getting a fresh instance, you can just install an MQTT broker, as we did on the Raspberry Pi, and access the broker through the IP address of your server. Please follow the same process given in *Chapter 1, Introduction to Raspberry Pi and MQTT* (the Raspberry Pi as an MQTT broker sub-section), to install the mosquitto open source MQTT broker on the virtual server.

  In this chapter, we will be choosing *Digital Ocean* as the platform of choice. We will set up a basic Linux server instance and install and demonstrate the use of our global MQTT broker through a simple project: we will send some messages from our Pi to the broker.

There are a lot of alternatives when it comes to choosing a platform for hosting a virtual instance. Some popular free sources that can be used are **Amazon Web Services (AWS EC2 Service)** and **Google Cloud Platform (GCP)**, which allow you to spawn a single virtual instance under their free-tier plan. But please keep in mind that if you are not cautious about staying within the limits of the free tier, you may end up getting billed heavily. Hence, I am using Digital Ocean for this tutorial.

The choice of setting up a virtual server has the advantage of providing the user with additional customization options and, at times, being more cost-efficient. But the downside of this is that you need some experience working with Linux to get the most out of all the available options. Hence, I would suggest that beginners stick with the first option.

In the upcoming subsections, we will cover all the steps required to set up your global MQTT broker in both cases and how to test that they are successfully up and running.

There will be two subsections – one describing the setup process when using an online MQTT broker service and another describing the setup process wherein we will get a Digital Ocean droplet (their term for an instance) and install the mosquitto MQTT broker on it.

Option 1 – online MQTT broker

In this subsection, we will set up an online MQTT broker and test it through the Node-RED interface of our Raspberry Pi. Then, we will discuss the advantages and disadvantages of this option.

I chose a free MQTT broker so that everyone of all skill levels can try this. We will set up a free cluster on HiveMQ Cloud, which is where our broker will be deployed. We can access it using the credentials and the hostname we obtain after the setup process. After setting up the broker, we will develop a simple Node-RED flow to test the broker:

Figure 7.4 – HiveMQ Cloud

Follow these steps:

1. HiveMQ Cloud is a cloud-native IoT messaging broker service provided by HiveMQ. There is a free tier available too, which lets you set up to 100 MQTT topic sessions.

Open the HiveMQ Cloud site by going to `https://www.hivemq.com/mqtt-cloud-broker/`:

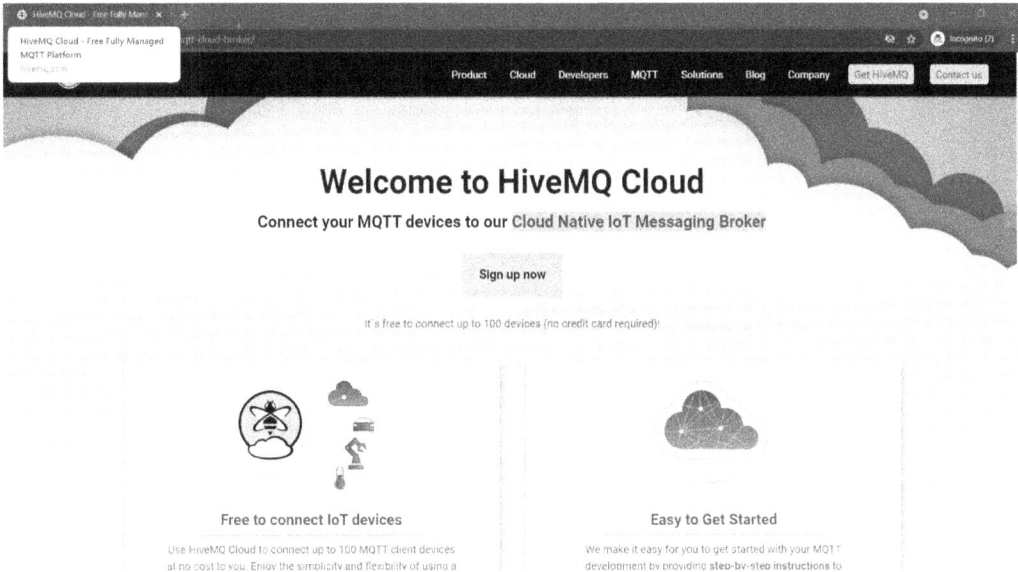

Figure 7.5 – Hive MQ Cloud home page

2.   From the home page, just press the **Sign up now** button. This will redirect you to the HiveMQ Cloud portal. From this page, you will be able to sign up for free:

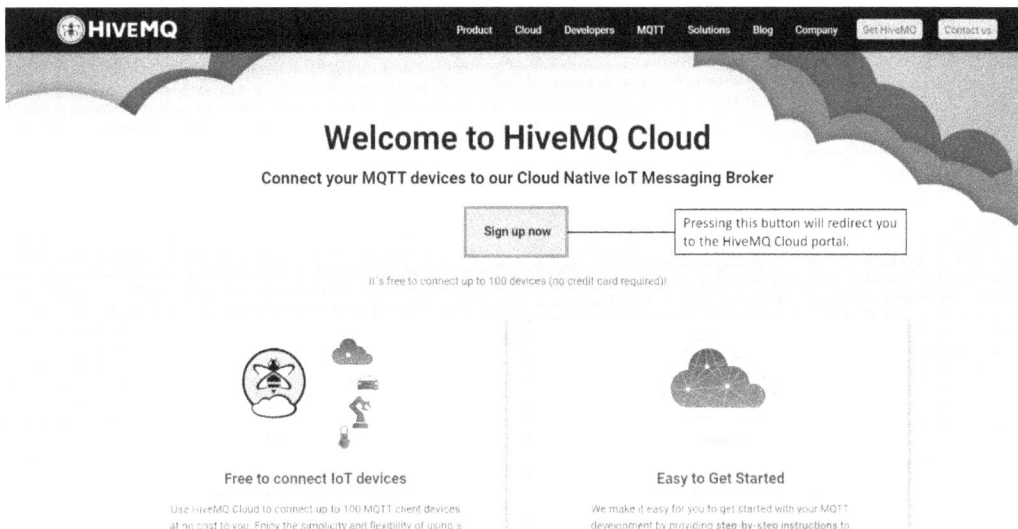

Figure 7.6 – Pressing the Sign up now button

3. Once you get to the Cloud portal, you will see a login page. Just go to the **Sign Up** tab by clicking on the **Sign Up** text. There, just enter the email and password as instructed (it would be wise to use a new email and not your personal email).

    Once you've done that, just click the **Sign Up** button to proceed to the next step:

Figure 7.7 – The HiveMQ Cloud portal to sign up

4. Next, a new page will open, where you will have to agree to the **Terms of Service** defined by the HiveMQ website. Just agree to these by checking the provided checkbox. Then, click the **Continue** button:

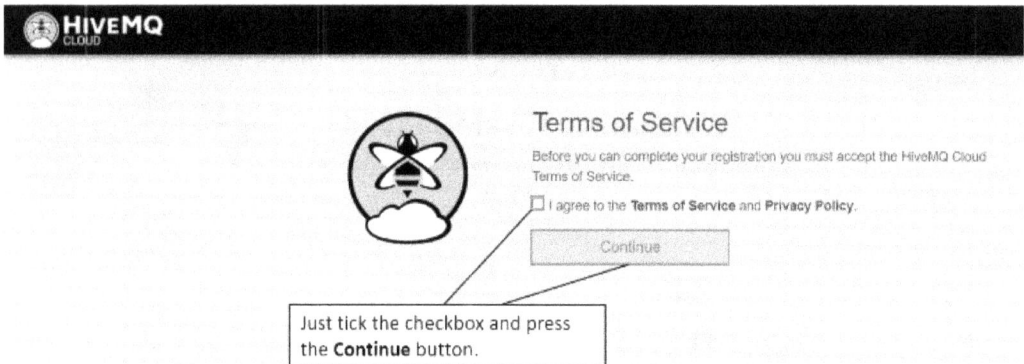

Figure 7.8 – Agreeing to the Terms of Service

5. Now, you will be redirected to a new page that will state that you need to verify your email address to continue (refer to *Figure 7.9* and *Figure 7.10*). A verification email will have been sent to the email ID that you provided while signing up.

Just open that email and click on the **Confirm my account** button. This will redirect you to the login page, which indicates that your email ID has been verified and that your account is active:

Welcome to HiveMQ Cloud.
Please verify your email address before using HiveMQ Cloud.

Please click the verification link in the email that has been sent to your inbox.
If you did not receive the verification email, please contact cloud@hivemq.com for help.

You will be automatically redirected to the login page.

Figure 7.9 – The email verification message

The verification email text will look something like this:

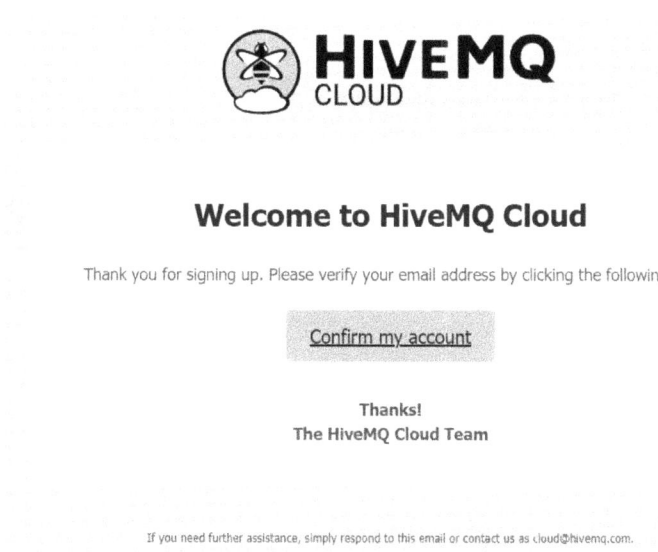

Welcome to HiveMQ Cloud

Thank you for signing up. Please verify your email address by clicking the following link

Confirm my account

Thanks!
The HiveMQ Cloud Team

If you need further assistance, simply respond to this email or contact us as cloud@hivemq.com.

Figure 7.10 – Verification email

6. After that, you will be redirected to the **Login** page. Just enter the required credentials (email and password) and click the **Login** button. You will be redirected to a new page where you will be asked for some additional information:

Thanks for signing up!

We'll just need a few more details and you'll be ready to go

First name   Last name

First name   Last name

Job Title

Job Title   ↕

Company

Company

Phone

Phone

All fields are required   Continue

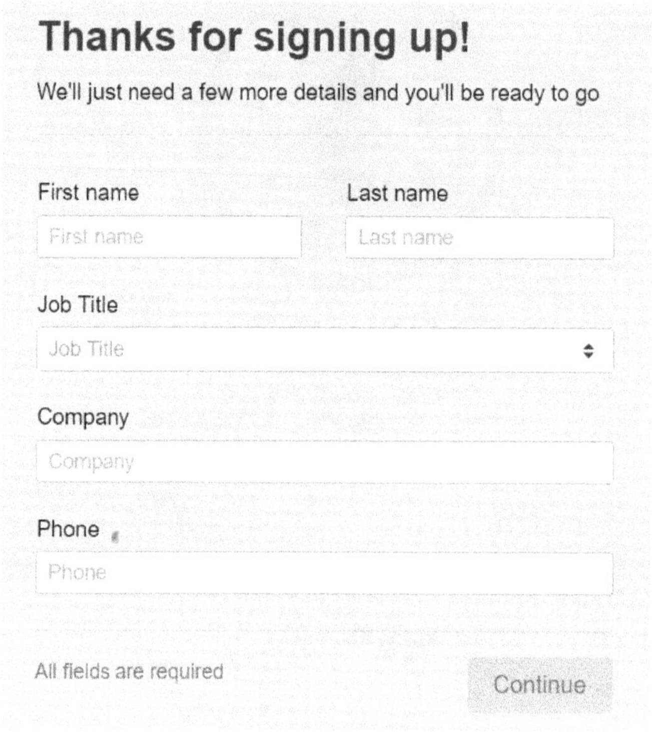

Figure 7.11 – Entering additional information

7. Once you have done this, click the **Continue** button. This completes the sign-up process. You will be redirected to a new page where you have to choose the cloud provider where your cluster will be located. At the time of writing, there are two options:

- **Amazon Web Services (AWS)**

- **Azure**

You can choose any provider of your choice. I have chosen AWS for the time being. The following screenshot shows what the web page will look like:

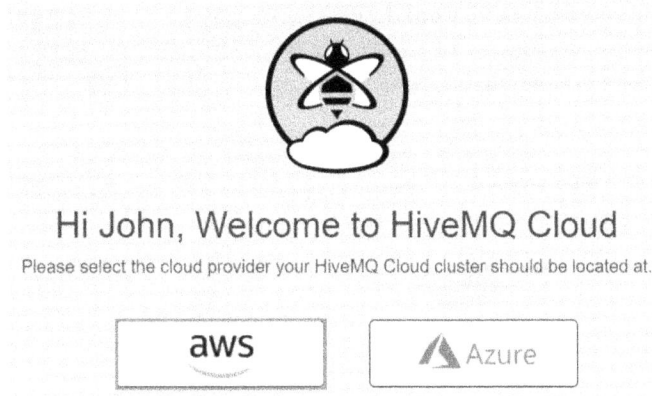

Figure 7.12 – Choosing a cloud provider

8. Once you have selected a provider, you will be redirected to the **Cluster Details** page – specifically, to the **Getting started** sections.

    You will have to just follow the steps shown on the screen. The first step would be to set up the credentials for your IoT devices. You will see the following page:

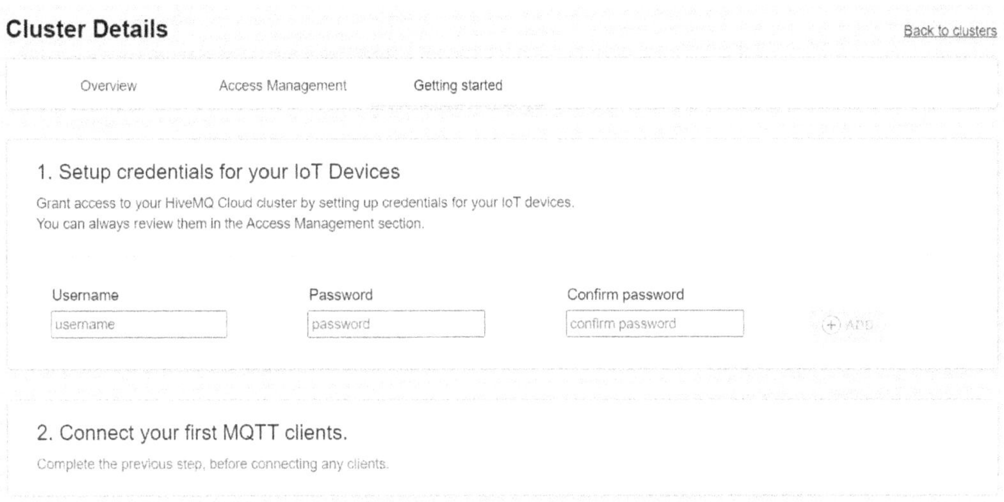

Figure 7.13 – The Cluster Details (Getting Started) page

9. The first step is to set up the MQTT broker credentials. These will be required when we try to initiate a connection from any client to the broker.

Just enter the **Username** and **Password** details of your choice. You will also need to re-enter the password in the **Confirm password** textbox. Once you've done that, just click the **Add** button. I have entered the following sample credentials:

- **Username**: `test-user`

- **Password**: `Demo@password123`

- **Confirm password**: `Demo@password123`

These can be seen in the following screenshot:

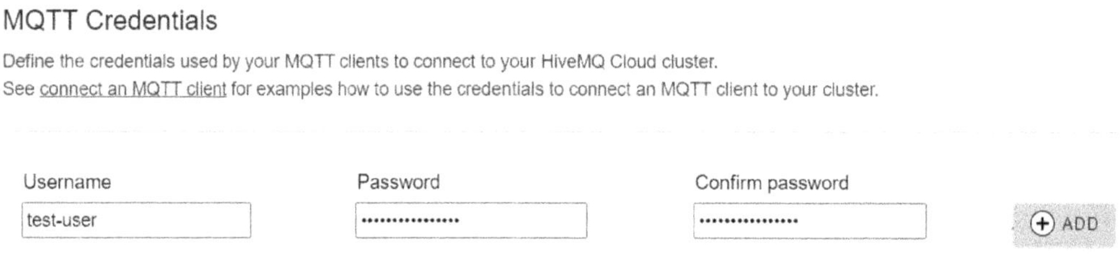

MQTT Credentials

Define the credentials used by your MQTT clients to connect to your HiveMQ Cloud cluster.
See connect an MQTT client for examples how to use the credentials to connect an MQTT client to your cluster.

Username	Password	Confirm password	
test-user	●●●●●●●●●●●●●●●	●●●●●●●●●●●●●●●	⊕ ADD

Figure 7.14 – Entering the MQTT credentials

10. Once the credentials have been added, a new section will open called **Connect your first MQTT clients.**

    Here, you will find tutorials on how to use all the available tools with your new broker, as well as how to define the necessary configurations when trying to connect to the same broker through code:

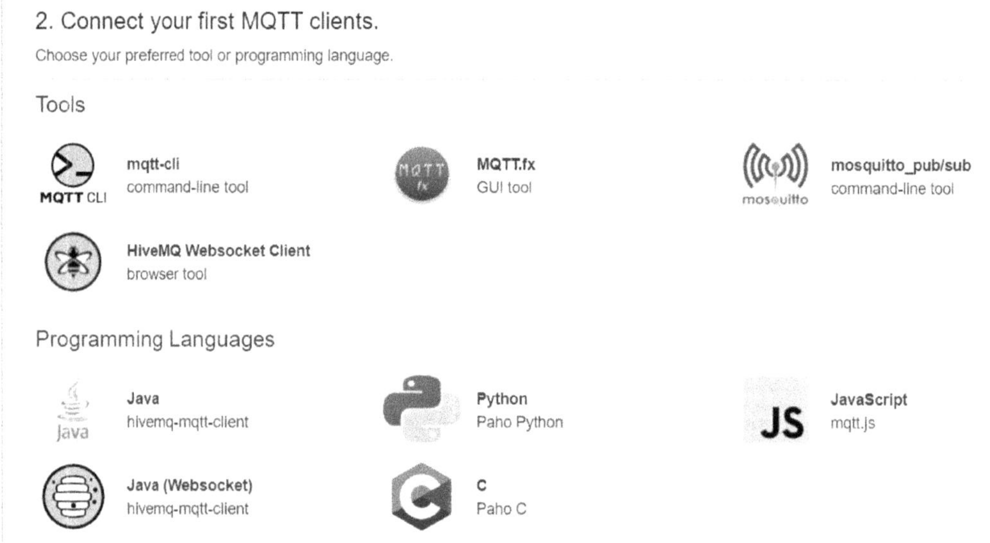

2. Connect your first MQTT clients.

Choose your preferred tool or programming language.

Tools

mqtt-cli
command-line tool
MQTT CLI

MQTT.fx
GUI tool

mosquitto_pub/sub
command-line tool
mosquitto

HiveMQ Websocket Client
browser tool

Programming Languages

Java
hivemq-mqtt-client
Java

Python
Paho Python

JavaScript
mqtt.js
JS

Java (Websocket)
hivemq-mqtt-client

C
Paho C

Figure 7.15 – Tutorials for different tools and programming languages

11. This completes the setup process. With that, you have finally set up your very own online MQTT broker. You can find the details about your cluster in the **Overview** tab.

Here, you will find the following details:

- Hostname and relevant ports

- Basic cluster information

- Capacity (for various properties)

You can manage the active credentials from the **Access Management** tab. The credentials we first entered will already be active. These can be removed and more credentials can be added too:

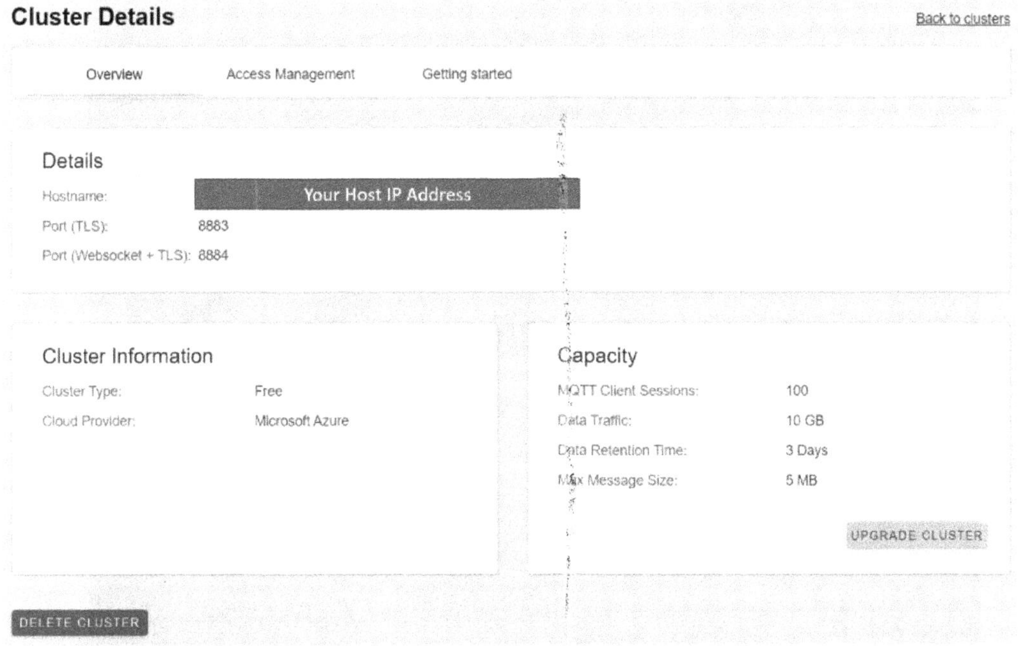

Cluster Details   Back to clusters

	Overview	Access Management	Getting started

Details

Hostname:	Your Host IP Address
Port (TLS):	8883
Port (Websocket + TLS):	8884

Cluster Information

Cluster Type:	Free
Cloud Provider:	Microsoft Azure

Capacity

MQTT Client Sessions:	100
Data Traffic:	10 GB
Data Retention Time:	3 Days
Max Message Size:	5 MB

UPGRADE CLUSTER

DELETE CLUSTER

Figure 7.16 – Cluster Details overview

Now that the setup process is complete, we will develop a simple Node-RED flow to demonstrate how to set up and use this MQTT broker hosted on the cloud.

Simple Node-RED flow to test the new broker

To create a simple flow to demonstrate how to work with this broker, follow these steps:

1. The first step is to start Node-RED after booting up your Raspberry Pi. Just open a new terminal on your Pi and type the following command:

```
node-red-start
```

You can also use the following command:

```
node-red
```

The output can be seen in the following screenshot:

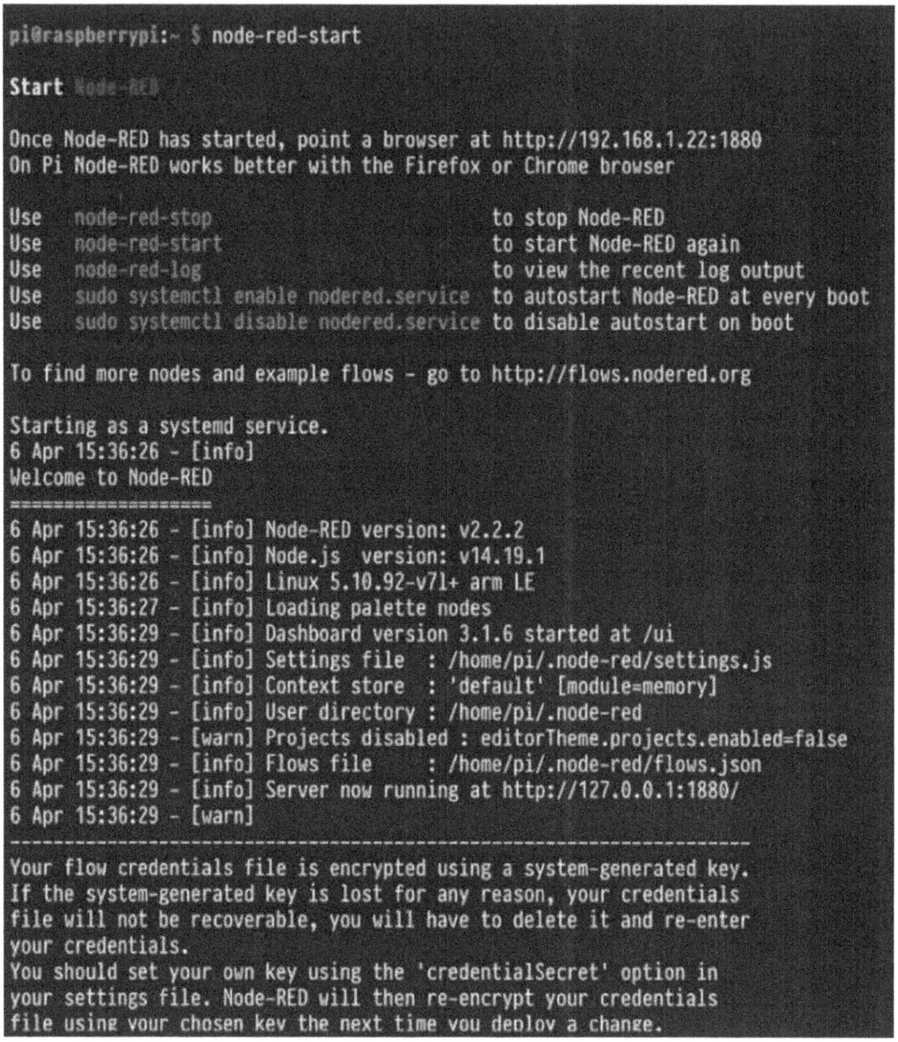

```
pi@raspberrypi:~ $ node-red-start

Start Node-RED

Once Node-RED has started, point a browser at http://192.168.1.22:1880
On Pi Node-RED works better with the Firefox or Chrome browser

Use    node-red-stop                            to stop Node-RED
Use    node-red-start                           to start Node-RED again
Use    node-red-log                             to view the recent log output
Use    sudo systemctl enable nodered.service    to autostart Node-RED at every boot
Use    sudo systemctl disable nodered.service   to disable autostart on boot

To find more nodes and example flows - go to http://flows.nodered.org

Starting as a systemd service.
6 Apr 15:36:26 - [info]
Welcome to Node-RED
===================
6 Apr 15:36:26 - [info] Node-RED version: v2.2.2
6 Apr 15:36:26 - [info] Node.js  version: v14.19.1
6 Apr 15:36:26 - [info] Linux 5.10.92-v7l+ arm LE
6 Apr 15:36:27 - [info] Loading palette nodes
6 Apr 15:36:29 - [info] Dashboard version 3.1.6 started at /ui
6 Apr 15:36:29 - [info] Settings file  : /home/pi/.node-red/settings.js
6 Apr 15:36:29 - [info] Context store  : 'default' [module=memory]
6 Apr 15:36:29 - [info] User directory : /home/pi/.node-red
6 Apr 15:36:29 - [warn] Projects disabled : editorTheme.projects.enabled=false
6 Apr 15:36:29 - [info] Flows file     : /home/pi/.node-red/flows.json
6 Apr 15:36:29 - [info] Server now running at http://127.0.0.1:1880/
6 Apr 15:36:29 - [warn]
------------------------------------------------------------------------
Your flow credentials file is encrypted using a system-generated key.
If the system-generated key is lost for any reason, your credentials
file will not be recoverable, you will have to delete it and re-enter
your credentials.
You should set your own key using the 'credentialSecret' option in
your settings file. Node-RED will then re-encrypt your credentials
file using your chosen key the next time you deploy a change.
```

Figure 7.17 – Starting Node-RED on the Pi

2. This command will start Node-RED and give you an IP address where you can access the Node-RED editor from any device connected to the same network as the Raspberry Pi.

    Open Node-RED on any browser of your choice by typing ß pi's ip address à:1880 into your address bar. This will open the Node-RED home screen. From there, just create a new flow by following the instructions provided to get a new, blank workspace:

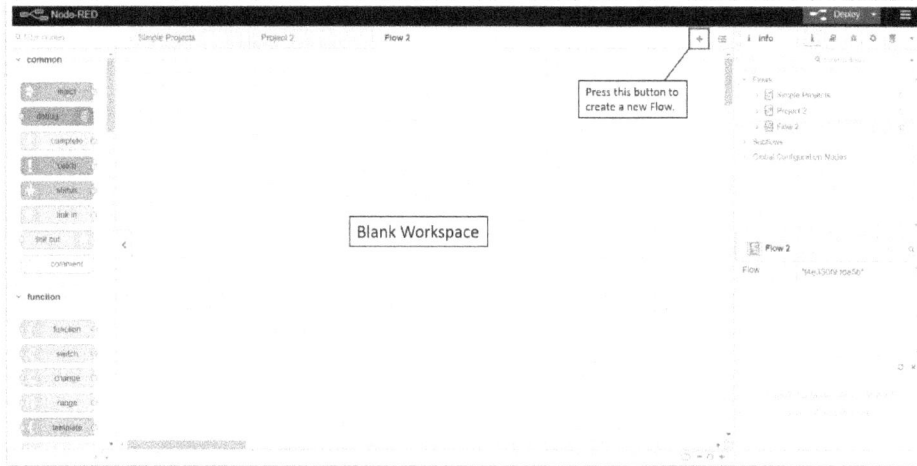

Figure 7.18 – Creating a new flow in Node-RED

3. Now, you just need to add the following nodes to your workspace:

    - One `Inject` node

    - One `Debug` node

    - One `MQTT in` node

    - One `MQTT out` node

    After adding these nodes, your workspace will look something like this:

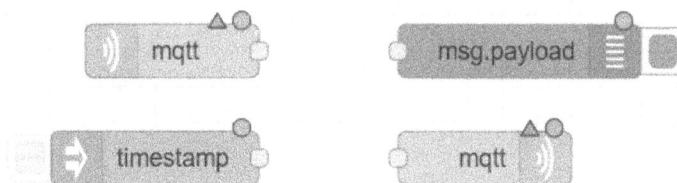

Figure 7.19 – The flow for testing our MQTT broker

Next, we must set up the nodes to function in a particular manner. For this project, we will do the following:

- When the `Inject` node is pressed, a *"Hello World!"* message is published on the **test/publish** topic.

- The MQTT in-node subscribes to the **test/subscribe** topic and whenever a message arrives on that topic, it prints the same on the **Debug** section (available on the right-hand side).

Please follow the instructions provided to set up each node (see *Figure 7.20*):

- `MQTT In Node`: Here, we just need to fill in the **Topic** textbox.
- `MQTT out Node`: Here, we just need to fill in the **Topic** textbox.
- `Debug Node`: No setup is required for the debug node.
- `Inject Node`: Just cross out the `msg.topic` option and for `msg.payload`, change the output option to `String` and fill in the text *"Hello World!"*.

In addition, we need to configure a new broker (our HiveMQ MQTT broker). We will need the following information from the HiveMQ cluster:

- Host IP address (hostname)
- MQTT credentials username
- MQTT credentials password

Refer to the following screenshot to see how to setup all the nodes for this project's flow.

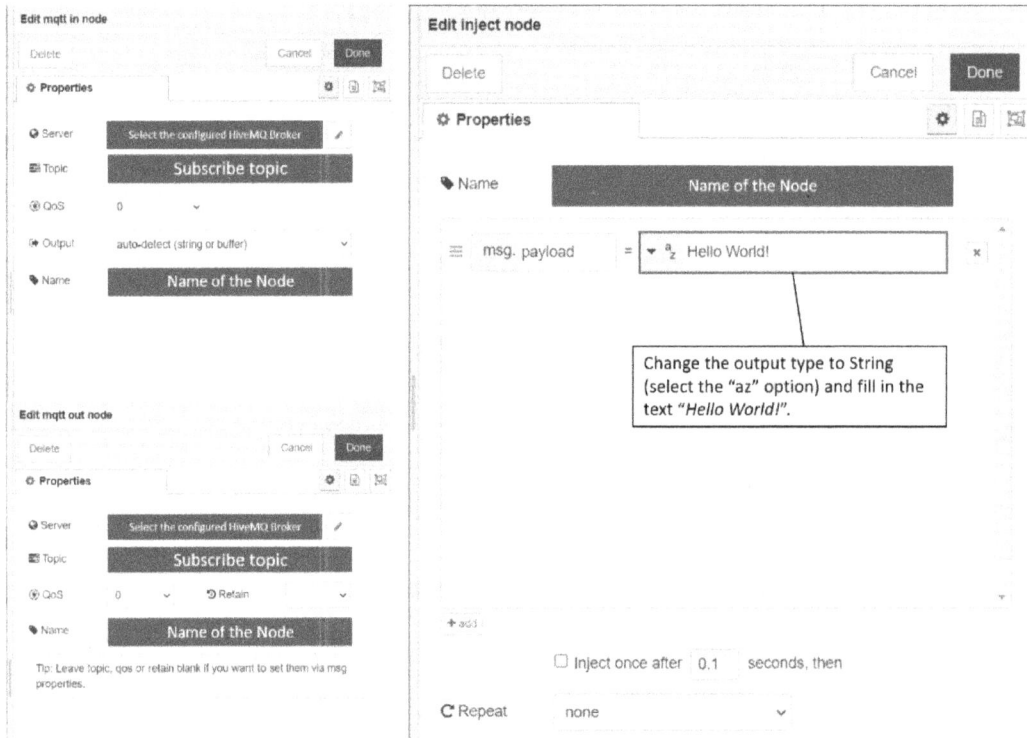

Figure 7.20 – All node configurations for this flow

Please refer to *Figure 7.21* to set up a new broker in Node-RED:

MQTT Broker Setup

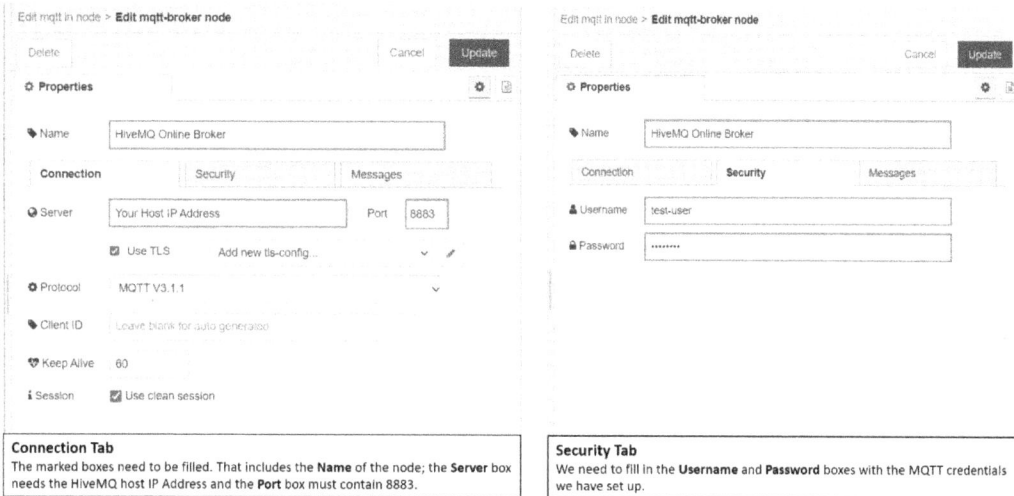

Connection Tab	Security Tab
The marked boxes need to be filled. That includes the **Name** of the node; the **Server** box needs the HiveMQ host IP Address and the **Port** box must contain 8883.	We need to fill in the **Username** and **Password** boxes with the MQTT credentials we have set up.

Figure 7.21 – MQTT broker setup

Once all the nodes have been set up, you are ready to create your final flow. The following screenshot shows what the final flow for this project looks like:

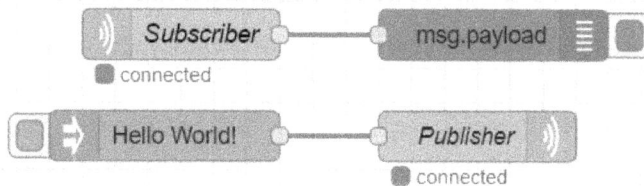

Figure 7.22 – Final project flow

Now that the project flow is complete, we will see this this project in action!

Project demonstration

For project demonstration purposes, we will use HiveMQ's MQTT WebSocket client. The following screenshot shows what the client looks like:

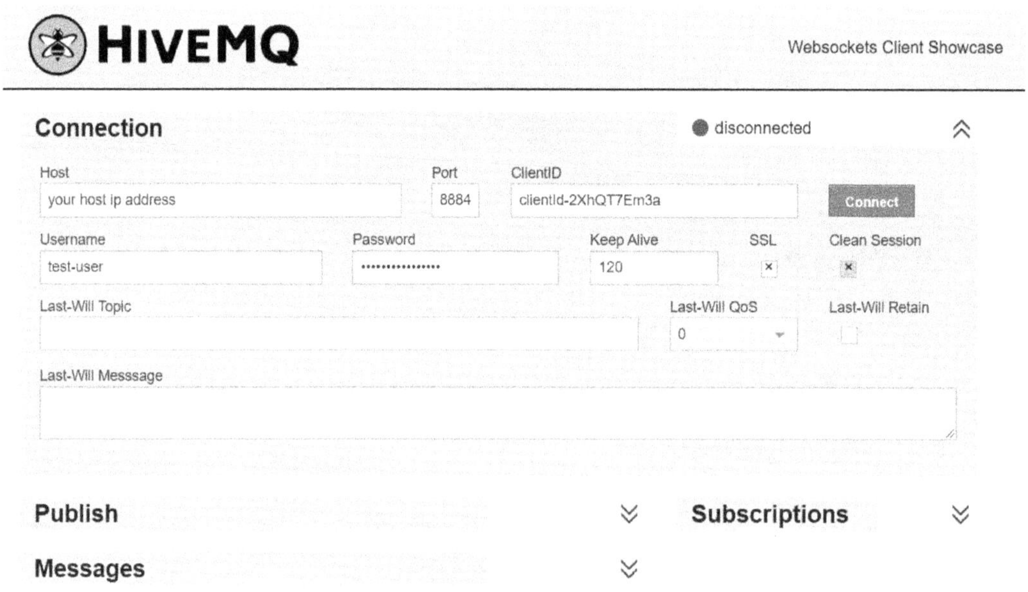

Figure 7.23 – HiveMQ's MQTT WebSocket client

This tool lets you connect to your broker and then, via its interface, publish and subscribe messages to particular topics. Here, we will use the topics we configured into our Node-RED flow.

The first thing you must do is deploy the flow on Node-RED. Once you've done that, just open the HiveMQ WebSocket client by going to `http://www.hivemq.com/demos/websocket-client/`.

First, you will need to initiate a connection to your broker. That can be done via the **Connection** tab on the WebSocket client's page. After opening the page using the aforementioned link, you just need to fill in all the necessary fields required to connect to your MQTT broker:

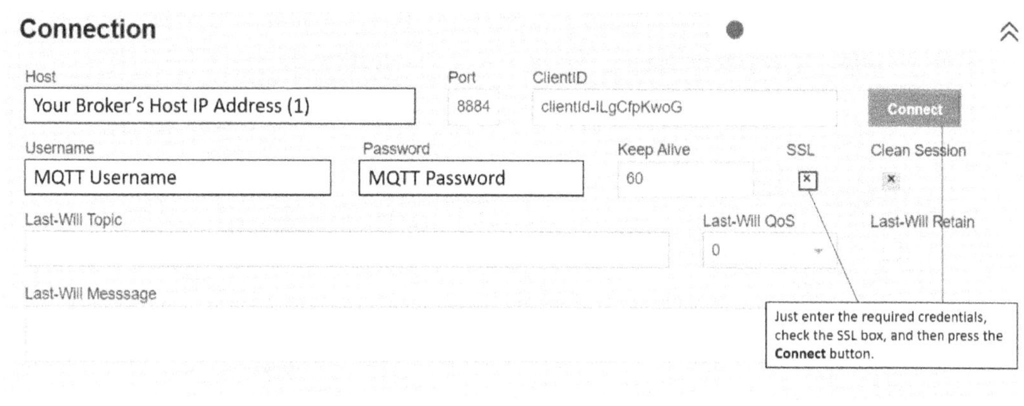

Figure 7.24 – Establishing a connection to the HiveMQ MQTT broker

Once the connection has been established, we are ready to proceed with the demonstration. Next, we need to do the following:

- Subscribe to the **test/publish** topic. This will let us monitor any messages or payloads that arrive on this particular topic. We have configured the node to send the *"Hello World!"* message to this topic when the **Inject** node is triggered.

- Publish the *"Hello World!"* string on the **test/subscribe** topic. We have subscribed to this topic in our Node-RED flow. So, whenever a message is published on this topic, the payload is printed on the **Debug** tab in Node-RED.

You must set up the WebSocket client so that it can do this. To do this, just follow the instructions provided in the following screenshot:

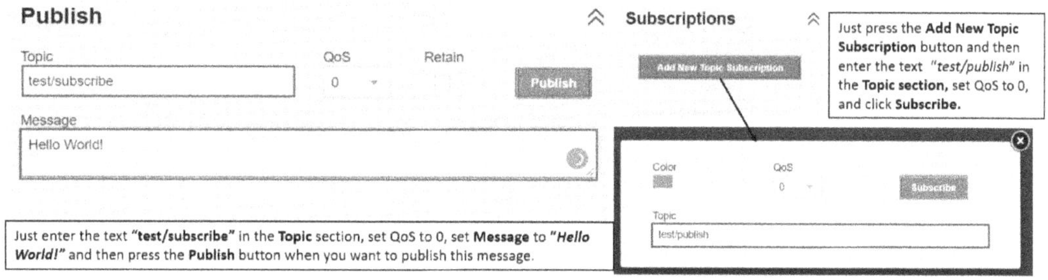

Figure 7.25 – MQTT WebSocket client setup

Now, we are ready to see the final project demonstration. This project will do the following:

- Whenever we trigger the inject node of our project, a *"Hello World!"* message will be printed in the **Messages** section of the WebSocket client.

- When you publish a message to the **test/subscribe** topic through the WebSocket client, the same message will be printed on the **Debug** tab in Node-RED.

The following screenshot shows the output of the publish message:

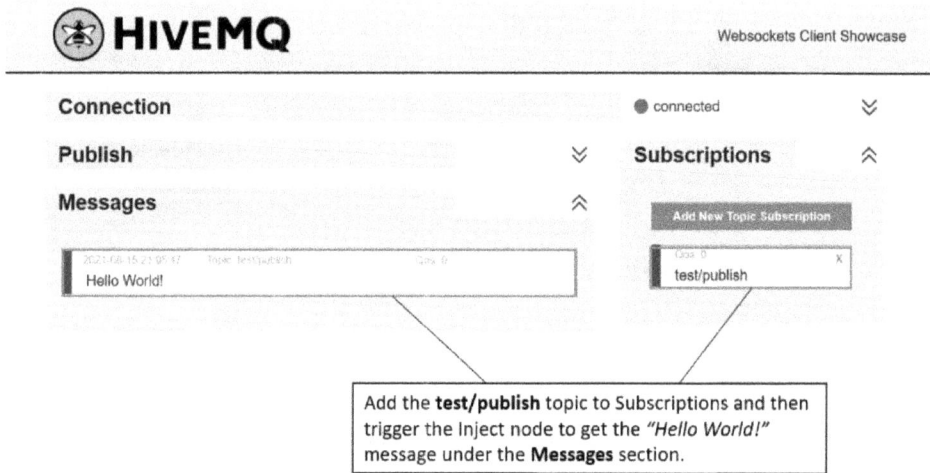

Figure 7.26 – Publish message output demonstration

The following screenshot shows the output of the subscribe message:

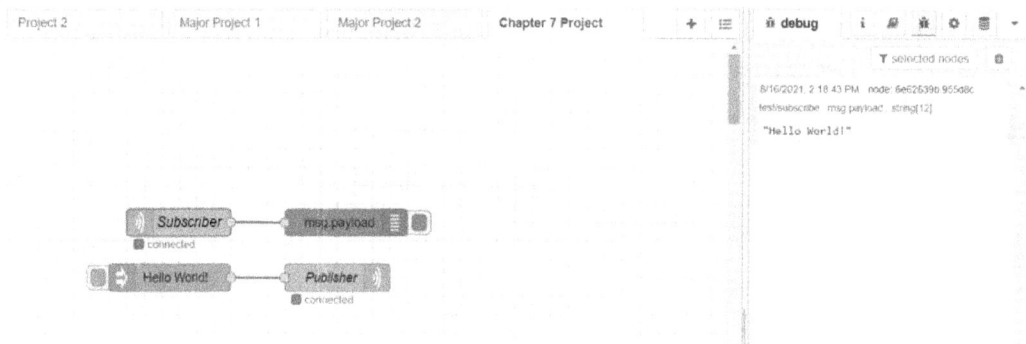

Figure 7.27 – Subscribe message output demonstration

With that, we have come to the end of this section. In the next section, we will discuss the second option we can adopt to make our MQTT broker global: hosting it on a virtual server. We will discuss its pros and cons and host our very own MQTT server on the Digital Ocean platform.

So, let's get started!

Option 2 – virtual server

In this section, we will discuss the second option that we have: hosting the MQTT broker on a virtual server. Several popular sites provide this service. However, for this project, we will choose Digital Ocean. (*Please keep in mind that you will need a credit card as this will cost you around 5 dollars a month.*)

As discussed earlier, there is a free-tier option available too. You can opt for a free trial on AWS or GCP. In both cases, you will be provided with free credits that you can use in any of the available services the platform has to offer.

> **Important Note**
>
> Even though the aforementioned options are free, there are certain conditions. First, you need to be a first-time user (or have an email ID that hasn't been registered). Second, you still need a credit card for verification.

The charges that are incurred on these platforms once the free-tier threshold is crossed are considerably more compared to Digital Ocean. However, using any of the two platforms is also acceptable. The only change in the tutorial would be the process of spawning your virtual server. There are several tutorials available on how to set up a virtual server on AWS or GCP.

In this subsection, we will walk through the process of setting up a new virtual server (called droplet) on Digital Ocean and then install and set up an MQTT broker on that server. Then, we will test the broker with the same project we created when we tested our HiveMQ's online MQTT broker.

So, let's begin with the setup:

1. **Creating a new Digital Ocean account**: You will need to create a Digital Ocean account to access the control panel and create a new droplet. To create a new account, navigate to the Digital Ocean new account registration page: `https://cloud.digitalocean.com/registrations/new`.

   You can choose to register via email, Google, or GitHub. Once you've confirmed your account, you will need to enter your credit card or PayPal information. This information is collected to verify your identity and keep spammers out. You will not be charged until you choose a plan and confirm your subscription, which we will cover in *Step 4*. You may see a temporary pre-authorization charge to verify the card, which will be reversed within a week.

   Once your information has been accepted, you will be taken to a window that says **Registration Complete**. You are now ready to proceed to the next step.

2. **Logging in to your Digital Ocean account**: Once your account has been successfully created, just log in to your account; you will see a control panel.

   This is the home screen you will get for your account. This is where you can manage all the droplets, databases, and domains and also navigate through all the services Digital Ocean has to offer.

Moreover, you can create different projects so that you can keep track of the resources that are used in individual projects.

For this tutorial, we will be creating a simple virtual server instance known as a **Droplet**. The following screenshot shows what the control panel for your account will look like:

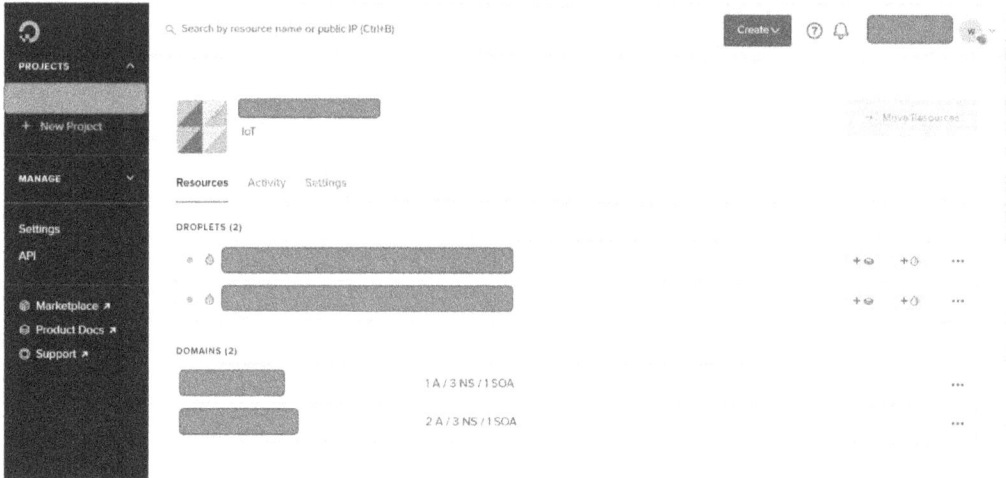

Figure 7.28 – My Digital Ocean control panel

3. **Setting up a new Droplet**: Click the **Create** button at the top right to expand the menu. There, click the **Droplets** option, which will redirect you to a new web page:

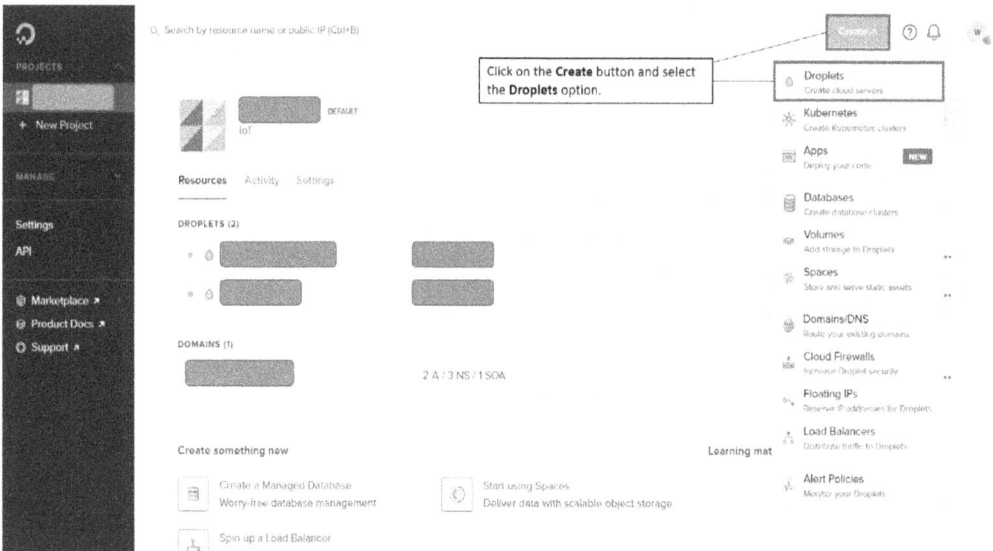

Figure 7.29 – Creating a new droplet

You can even open this web page from the home screen of the dashboard, provided that you don't have any active droplets, in which case you will have a **Get started with a Droplet** button in the **Resources** tab.

From this page, you can configure various droplet characteristics such as CPU, memory, OS, and more. The most used and popular configuration options will be preselected, but we will change them according to our requirements.

4. **Choosing an image for your Droplet**: Now, we will choose which OS our droplet will have. Various OSs are available on this platform and they have been widely divided into four main categories:

    *   **Distributions** are basic or vanilla OS images such as Ubuntu and Fedora. They have no additional packages pre-installed.

    *   **Container distributions**, which include the Rancher OS.

    *   **Marketplace** images consist of preconfigured applications, such as WordPress, LAMP, or application-specific Docker images, to help simplify getting started. These are some popular images that are used for specific applications. They come with all the required software packages pre-installed. One such example is the popular LAMP server configuration, which we will be covering in the last chapter of this book (*Chapter 8, Project Prototype to Product – How?*).

    *   **Custom images** are required if the OS for your requirements is not available for any of the aforementioned options. In that case, you will have to upload your own OS image through this option so that it will be flashed on your droplet.

    For this tutorial, we will be using a basic **Ubuntu 20.04 64-bit LTS** distribution for our droplet. This is because it is the most used Linux OS with a very good developer community and support.

    Please note that this option will be chosen for you by default:

Create Droplets

Choose an image ?

Distributions   Container distributions   Marketplace   Custom images

Ubuntu	FreeBSD	Fedora	Debian	CentOS	Rocky Linux
20.04 (LTS) x64 ⌄	Select version ⌄	Select version ⌄	Select version ⌄	Select version ⌄	Select version ⌄

Figure 7.30 – Choosing an OS for your virtual server

5. **Choosing a plan**: In this section, you will have to choose a particular plan according to your compute and storage requirements, such as CPU, memory, storage, and more.

The DigitalOcean platform has segregated its plans into five basic categories:

- **Basic Droplets** is for most basic and low compute-intensive applications such as web hosting a single website. For **Basic Droplets**, you can also choose between Regular and Premium CPU configurations, based on your requirements.

- **General Purpose Performance Droplets** is an option that provides a balance between memory and CPU compute power.

- **CPU-Optimized Performance Droplets** are for applications that have CPU-intensive tasks such as batch processing of several pipelines or video processing tasks. They give you the best CPU performance but the memory options are limited.

- **Memory-Optimized Performance Droplets** are used when you need a lot of RAM in your application, which can be a database server with several concurrent query executions.

- **Storage-Optimized Droplets** are best suited for applications that need high-performance storage. One example can be maintaining a data warehouse.

For this project, we will choose the most basic configuration available as we only need to run an MQTT broker on the server for now. The plan is a **Basic Droplet** with the following:

- A shared regular Intel processor

- 1 GB of RAM and 1 CPU core

- 25 GB of SSD disk space

- A 1,000 GB data transfer limit per month (this is a very important feature only available in Digital Ocean. In AWS or GCP, you are charged extra for the data transfer costs.)

Now, let's see how this section looks on their website. As we can see, the plan will cost you **$5** per month:

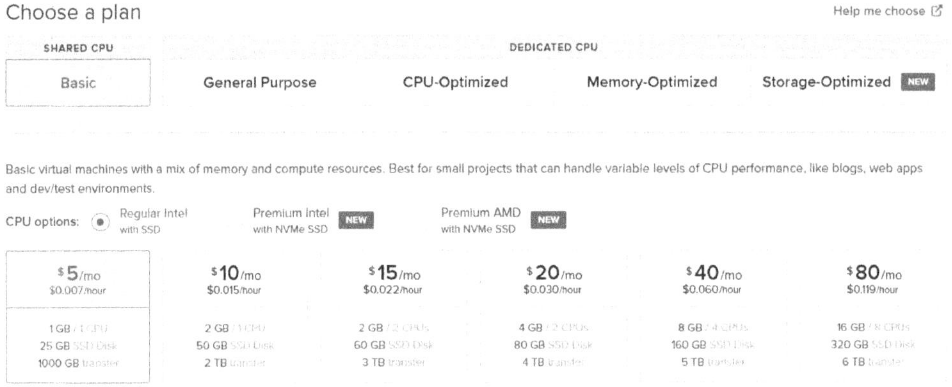

Figure 7.31 – Choosing the most basic plan

6. **Choosing a data center region**: In this section, we must select a region in which our droplet will be created.

The region with the best performance and minimum latency has already been selected but you can choose the data center nearest to you and your users. Several configurations are only available in a specific region, so you should plan for that as well:

Choose a datacenter region

Figure 7.32 – Choosing an appropriate data center region

The web page grays out any non-supported configurations by default, so you should not have any issue selecting a region for your droplet.

For this tutorial, you can choose any data center of your choice. I will just keep the default center selected. Please note that if you wish to upgrade your plan in the future, you may need to change your data center region according to the plan you choose as not all configurations are available in every region.

7. **Authentication**: In the **Authentication** section, you can choose what type of security option you would like for your droplet. There are two options provided by Digital Ocean:

    -   **SSH keys**, in which you will have to upload your own private key so that it can be used by the server when you try to remotely access it using the same key.

    -   **Password**, which is the most basic form of authentication, wherein the system will let you set up a password that you will be asked for every time you log in to your droplet.

    These options can be seen in the following screenshot:

Authentication ?

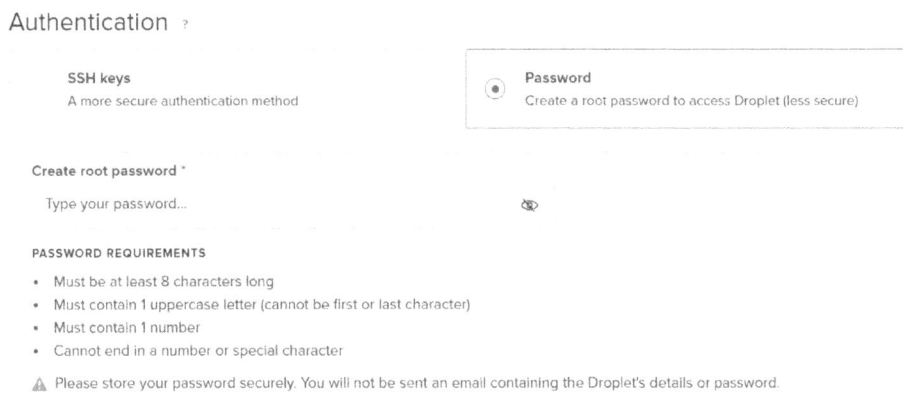

Figure 7.33 – Choosing your preferred authentication method

If you choose to use SSH keys, you will need to upload a key or select an existing key that has been previously uploaded for this Digital Ocean account. As we are doing this for the first time, we will need to upload a key.

You can even add multiple keys for authentication. Click on the **New SSH Key** button to upload a new key. If you wish to use SSH keys for authentication, please refer to the following reference link, which is a tutorial that walks you through the process of creating your own set of SSH keys:

`https://docs.digitalocean.com/products/droplets/how-to/add-ssh-keys/`

For this project, we will use a simple password for authentication. Just choose the **Password** option and then type in a password of your choice (keep the requirements listed in *Figure 7.33* in mind). This will create a password for the root user, which you can use to access your droplet.

8. **Reviewing the options and creating the Droplet**: Once everything is done, there is a **Finalize and Create** option, wherein you can choose how many droplets with the aforementioned configurations you want to create and what the hostname of each droplet (a default name is provided) should be. Also, you can choose in what project you want to create this droplet and specify some tags for the droplet too, which allows you to organize and relate between multiple droplets:

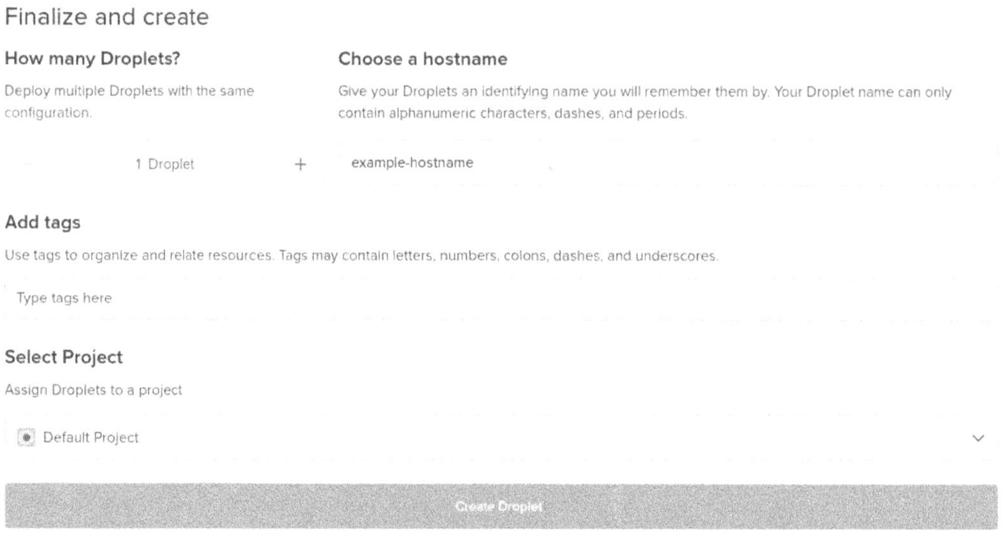

Finalize and create

How many Droplets?

Deploy multiple Droplets with the same configuration.

Choose a hostname

Give your Droplets an identifying name you will remember them by. Your Droplet name can only contain alphanumeric characters, dashes, and periods.

1 Droplet   +   example-hostname

Add tags

Use tags to organize and relate resources. Tags may contain letters, numbers, colons, dashes, and underscores.

Type tags here

Select Project

Assign Droplets to a project

⦿ Default Project

Create Droplet

Figure 7.34 – The Finalize and create section for droplet creation

Once you have selected your configurations, just click on the **Create Droplet** button to start the creation process. A progress bar will appear, showing how close your Droplet is to being ready:

DROPLETS (1)

◇ ubuntu-s-1vcpu-1gb-sfo2-01

Figure 7.35 – Droplet creation progress bar

Once this process is complete, you will be able to see your droplet listed on the home screen, along with a green dot to the right and the IP address of the droplet, which can be used to access it remotely:

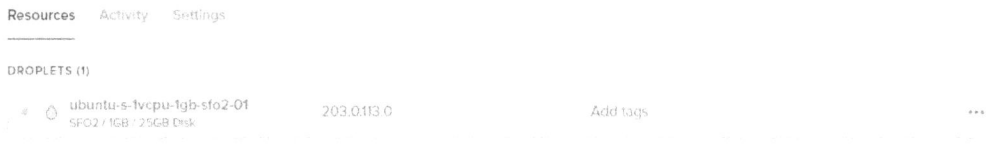

Resources   Activity   Settings

DROPLETS (1)

◇ ubuntu-s-1vcpu-1gb-sfo2-01   203.0.113.0   Add tags   ...
  SFO2 / 1GB / 25GB Disk

Figure 7.36 – The Resources section after droplet creation

To learn more about the droplet, just click on the droplet's name. This will open a new page with all the necessary details and usage analytics graphs.

Congratulations! You have successfully configured and created your first Digital Ocean droplet. Now, we will set up an MQTT broker on this droplet and test that it's working using the same project we created earlier.

MQTT broker setup

Setting up an MQTT broker on the droplet is a very easy process. We essentially have to follow the same steps that we did when we set up the Mosquitto MQTT broker on the Raspberry Pi.

To do that, we will need access to the terminal of the droplet we created. Here, we have two options. Let's look at the first:

1. We can access it through the **Web Console** area provided by Digital Ocean. To access this console, just click on your droplet from the control panel and then click on the **Console** button, as shown in *Figure 7.37*.

2. You will be asked to enable the **Droplet Console** area. Just follow the instructions to install a new console agent. Once this is done, you will be able to access your droplet through the web console:

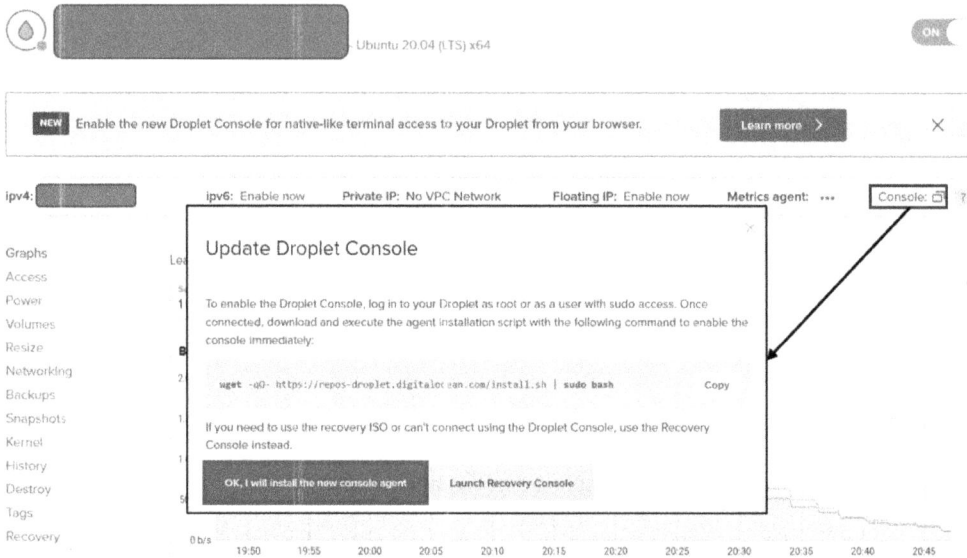

Figure 7.37 – How to access the web console to access your droplet

However, there are limitations while using the web console, including limited functionality and a lack of ease of access. Hence, we will opt for the second option.

Now, let's look at the second option:

1. We can use a third-party SSH client to access our droplet. Here, we will use **PuTTY**. Please note that this is only required for Windows systems. If you are using Linux or Mac, you will be able to SSH into any servers through the terminal itself. Moreover, once you install PuTTY, you will be able to do the same through your Windows command prompt:

Figure 7.38 – PuTTY for Windows

To install PuTTY on your system, go to `https://www.putty.org/`.

After the installation is complete, just open PuTTY on your Windows PC or laptop. Then, under **Host name**, enter the IP address of your droplet. This should open a terminal that will prompt you to enter a username and password. Just enter the following:

- **Username**: `root`

- **Password**: `<-droplet password->`

If you are using a Linux or macOS system, then simply open your terminal and enter the following command:

```
ssh root@<-droplet ip->
```

After entering this, you will be prompted to enter the password. Just enter your droplet's password and you will successfully log in to your server.

2. Once you have access to the terminal, you just have to type in the following command to install all the packages required to run an MQTT broker on this server (this is the same command we ran on the Raspberry Pi):

```
sudo apt install mosquitto mosquitto-clients
```

You may have to enter your password the first time you use `sudo`.

This will install all the required packages for your droplet. Once the installation is complete, we are ready to proceed to the next step, which is to enable our broker.

As was the case for the Raspberry Pi, we do not require the `mosquitto-clients` package to run the broker. However, with this package, we can emulate MQTT clients from the terminal itself, as well as on the server, which is great for debugging and testing.

3. Enable the broker using the following command:

```
sudo systemctl enable mosquitto
```

The broker should now be running. You can confirm this by running the following command:

```
sudo systemctl status mosquitto
```

The following screenshot shows the preceding command's output:

```
● mosquitto.service - Mosquitto MQTT v3.1/v3.1.1 Broker
   Loaded: loaded (/lib/systemd/system/mosquitto.service; enabled; vendor preset:
enabled)
   Active: active (running) since Tue 2021-03-16 16:33:30 IST; 3min 39s ago
     Docs: man:mosquitto.conf(5)
           man:mosquitto(8)
 Main PID: 2607 (mosquitto)
    Tasks: 1 (limit: 2062)
   CGroup: /system.slice/mosquitto.service
           └─2607 /usr/sbin/mosquitto -c /etc/mosquitto/mosquitto.conf

Mar 16 16:33:30 raspberrypi systemd[1]: Starting Mosquitto MQTT v3.1/v3.1.1 Broke
r...
Mar 16 16:33:30 raspberrypi systemd[1]: Started Mosquitto MQTT v3.1/v3.1.1 Broke
r.
```

Figure 7.39 – Output of the status command for the broker

The most important thing is that the **Active:** option should show an *active (running)* status, which will verify that our broker is up and running. If the status shows that your process is dead, you can simply restart your broker using the following command:

```
sudo service mosquitto restart
```

This will resurrect the process and should change the status to *running* again.

Great! You have successfully set up your very first fully customizable online MQTT broker running your very own server. Now, it is time to get our hands dirty with some coding to test our broker.

Testing our broker

To test this broker, we will use the same flow we created to test HiveMQ's online MQTT broker as the functionality to test remains the same.

The flow is shown in the following screenshot:

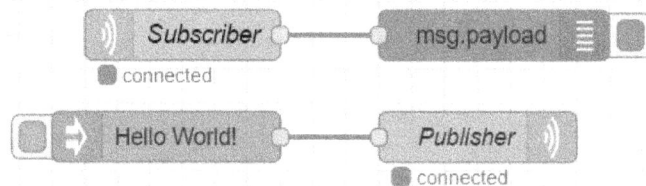

Figure 7.40 – Project flow used to test the HiveMQ broker

We will also use our server for testing purposes as we already have the client's package set up there.

The next step is to set up a new broker for our Node-RED flow. Just follow the instructions provided in *Figure 7.41* to do so. Please note that no username and password have been set up for the broker, so we can access it directly. This is not at all recommended as it raises a lot of security concerns.

It is very easy to set up a username and password combination for your broker – you just need to run a single command and change a few lines in a configuration file. Follow these steps:

1. Open a new terminal on your Pi and type the following command:

   ```
   sudo mosquitto_passwd -c /etc/mosquitto/passwd <USERNAME>
   ```

2. Type in the username of your choice instead of USERNAME. Then, the system will prompt you to enter a password, which will not show on your screen. Just type the password of your choice and press *Enter*.

3. This will create a password file named passwd in the /etc/mosquitto directory. All we need to do now is add some lines in the Mosquitto configuration file. To do this, just type the following command in the terminal:

   ```
   sudo nano /etc/mosquitto/mosquitto.conf
   ```

4. This will open the configuration file in the nano text editor. Go to the end of the file and add the following lines:

   ```
   allow_anonymous false
   password_file /etc/mosquitto/credentials
   ```

5. Once you have added these lines, press *Ctrl + X* on your keyboard and press *Y* to save the changes we just made.

6. Now, the final step is to restart our MQTT broker so that these changes can take effect. To do this, type the following command:

   ```
   sudo systemctl restart mosquitto
   ```

That is all you need to do to set up basic authentication for your MQTT Broker. However, since we are just testing the functionality, we will not be setting up a username and password for our broker:

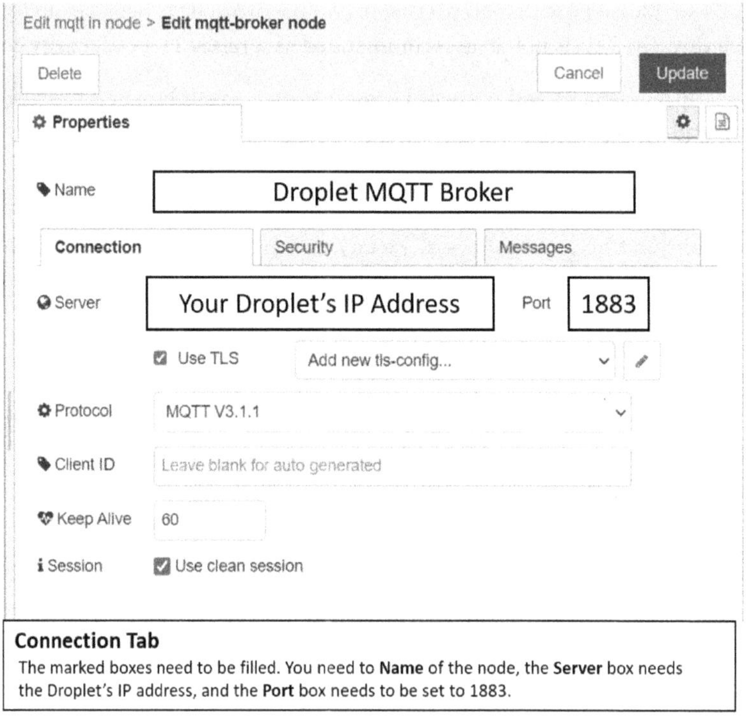

Figure 7.41 – Adding our new MQTT broker to Node-RED

Now, just change the broker settings for both the MQTT in and out nodes and redeploy the flow.

Next, we will need the terminal of our server again. We will be using the `mosquitto_pub` and `mosquitto_sub` commands to test whether our broker is up and running. `mosquitto_pub` and `mosquitto_sub` are command-line tools provided by Mosquitto for publishing and subscribing to topics using any MQTT broker. We used these when we tested the broker on our Raspberry Pi. We will follow the same steps to test this broker as well.

Type the following command on your terminal. The purpose of this command is to publish the message *"Hello World!"* on the **test/subscribe** topic:

```
mosquitto_pub -h <droplet ip> -p 1883 -t test/subscribe -m
"Hello World!"
```

Once you run this command, you should see that the same message is printed on the **Debug** tab of Node-RED:

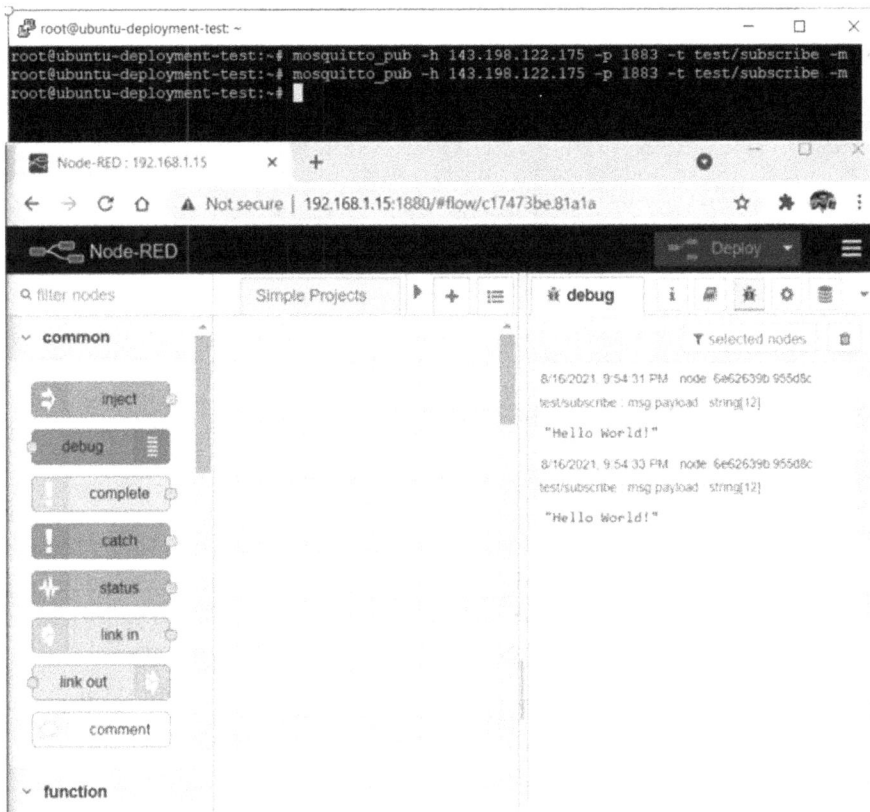

Figure 7.42 – Testing our broker hosted on a virtual machine using the mosquitto_pub command

> **Note**
> Please do not try to access the IP address shown in the preceding screenshot as it is no longer active.

This shows that we can now use our broker and publish a message on any topic through a client device connected to the internet (our Pi is the client in this case).

Next, we will check the *subscribe capabilities* through the mosquitto_sub command. Just type the following command on your terminal and press *Enter*:

```
mosquitto_sub -h <droplet ip> -p 1883 -t test/publish
```

This command will subscribe to the **test/publish** command. Now, we have configured our Node-RED flow to print a *"Hello World!"* message when we trigger the inject node.

We will do this by pressing the blue button on the inject node. This should show the same message printed on the terminal too:

```
root@ubuntu-deployment-test:~#
root@ubuntu-deployment-test:~# mosquitto_sub -h 143.198.122.175 -p 1883 -t test/publish
Hello World!
Hello World!
```

Figure 7.43 – Testing our broker hosted on a virtual machine using the mosquitto_sub command

Now, we can subscribe to any topic and access the arriving messages through a client device connected to the internet and our MQTT broker.

Summary

We covered a lot of interesting topics in this chapter. So, let's go through them once more to refresh our memory. We started by discussing the limitations of our present Raspberry Pi MQTT broker and then looked at the advantages of an MQTT broker, which can be hosted and accessed through the internet. Then, we discussed two options that can help us take our MQTT broker global and learned how to set up each and test them using a simple Node-RED flow on our Raspberry Pi.

We are almost at the finish line! We have covered a lot of content throughout this book. In the next chapter, we will cover the topics that will assist you in answering a very important question: *What next?*

8

Project Prototype to Product – How?

We are almost at the finish line. This book has covered all the essentials required to get yourself familiar with numerous concepts related to Raspberry Pi and MQTT. But this is just the beginning! Now, the most important question arises: *What Next?*

This chapter provides the answer to this question. We will be covering a lot of important things that will assist you in choosing your future path.

In this chapter, we will cover the following topics:

- Innovative project ideas
- IoT services provided by enterprise cloud platforms
- How to scale your projects using the current hardware

So, let's get started!

Innovative project ideas

"Building projects is the best way to learn a technology, period."

This is the motto I believe in. Learning a technology is not just about theory. You can read more than 100 books on a single piece of technology but to master it, you have to have hands-on experience in working with that technology. This helps you not only understand the nitty gritty details but also learn new things that you have to pick up while building a project:

Figure 8.1 – It all starts with an idea!

The projects chosen in this book were handpicked by me as I built them when I was first getting familiar with Raspberry Pi. We will cover two very popular project ideas that you can build to further expand your knowledge of Pi and the MQTT communication protocol.

Please note that this section only provides an overview of these ideas; you will have to research yourself and build them, which is how you will learn and master various technological skills.

Idea 1 – Home automation system

Smart homes are quickly gaining popularity, all thanks to the cheaper and more powerful hardware available on the market currently. This is not a new concept and has been implemented by thousands of DIY enthusiasts, including myself:

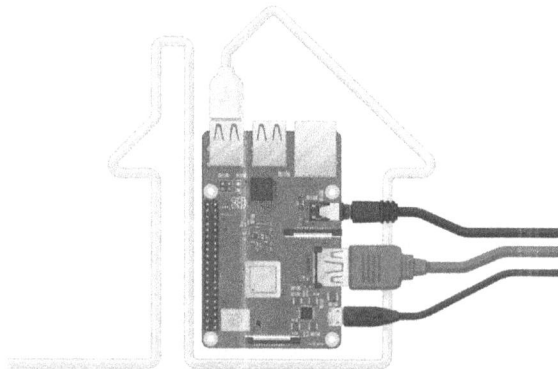

Figure 8.2 – Home automation in a nutshell

The preceding figure signifies how easy home automation is to implement. A single Raspberry Pi can act as a hub for your entire home automation system!

The main advantage of this idea is that you can implement it at any scale, from a single appliance to a whole house being controlled through a single hub. Let's discuss how you can use your existing skillset to create this project. First, by now, it should be clear how the hardware components will be used. You will use the following:

- A Raspberry Pi, which will act as the central hub for your home automation ecosystem. It will be used to control all the client or end node devices.

- The ESP32 and ESP8266-based home automation projects will be used as the client. You can use the Major Project 2 that you built directly for this. You will need multiple projects to create this at a larger scale. For instance, one such project will be required for each room of your apartment or house.

Idea 2 – air quality monitoring system

Air is essential to sustain life on our planet. As we all must have studied in school, air is 78% nitrogen, 21% oxygen, and 1% argon; the rest of the gases are very minute in volume. Nonetheless, several gases are categorized as polluters, with carbon dioxide being the main culprit. Apart from this, there are other minute solid particles that greatly affect human health, known as particulate matter. They are categorized based on particulate size (in microns), and the two major types are PM2.5 and PM10. Three main components can be used to monitor these with a conventional air quality monitoring system.

You can simply connect one of the commercially available sensors to your Raspberry Pi and write some simple Python code to start monitoring these parameters. You can choose any sensor of your choice for this purpose. Some options are as follows:

- **Honeywell HPM Particle sensor**: This provides laser-based light scattering particle sensing, including PM1.0, PM2.5, PM4.0, and PM10. It responds in less than 6 seconds, has UART, and has standard and compact versions. These are just some of the characteristics of this sensor:

Figure 8.3 – Honeywell HPM Particle Sensor

- **SDS011 sensor**: This is a recently developed air quality sensor created by a company called Nova Fitness. You can monitor PM2.5 and PM10 readings using this sensor:

Figure 8.4 – SDS011 sensor

You will also require a UART interface if you want to connect any of these sensors to the Pi. However, most sensors come with a USB to Serial connector, as shown here:

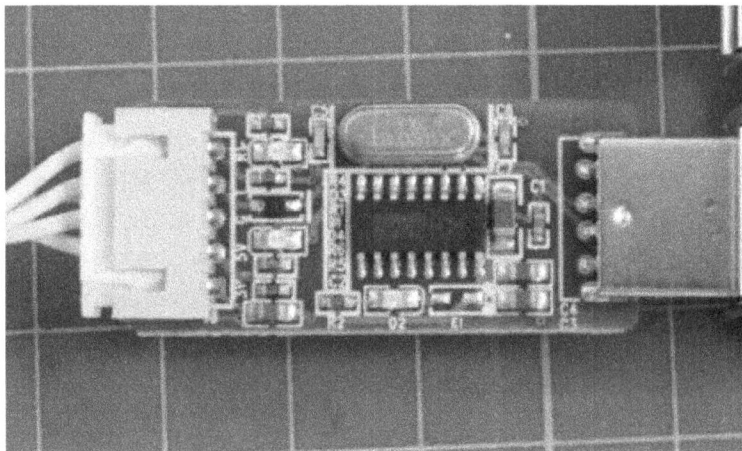

Figure 8.5 – USB to Serial connector for sensors

Regarding the software, you will need to code in Python to read the values from the sensor through UART. What you do with these values is up to you. You can create a Node-RED dashboard, like the one we created for the weather station, or you can use external data visualization or storage tools such as **Adafruit IO**, **ThingSpeak**, and so on. The following is an example dashboard for this project:

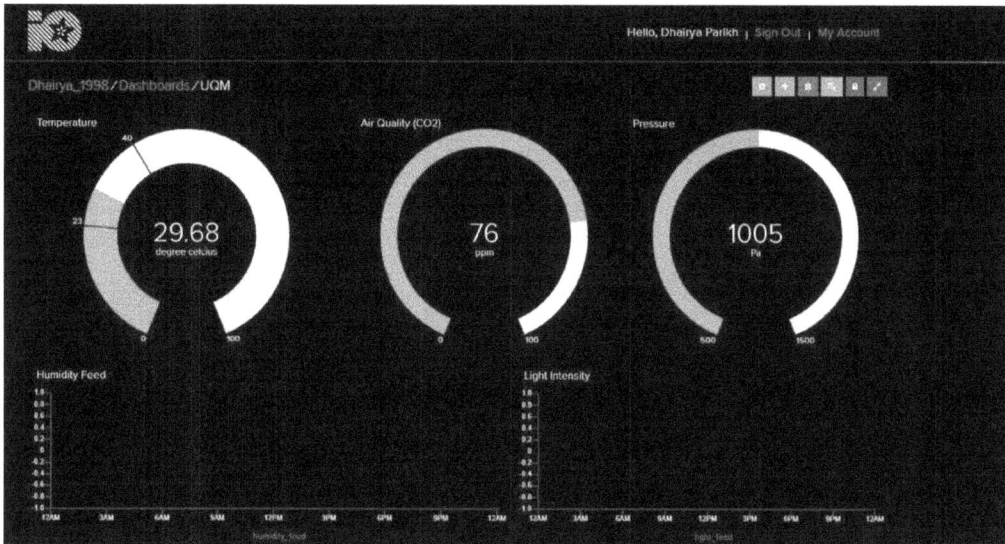

Figure 8.6 – Example quality monitoring dashboard (Adafruit IO)

In this section, we discussed two very popular ideas that you can implement using the skills you've learned throughout this book. Now, we will explore some IoT services offered by enterprise cloud platforms such as **Amazon Web Services** (**AWS**) and **Google Cloud Platform** (**GCP**) and create a simple project using one such service – you learn best by doing!

IoT services provided by enterprise cloud platforms

Modern cloud platforms such as AWS, GCP, and Azure provide a fully managed IoT system architecture that can help you do the following:

- Data collection and analysis
- Remote access and deployments
- IoT data security
- Third-party application integrations

The main advantage of using such a platform is that the hardware is completely managed by the cloud service provider, along with data security and access control management. This creates a very powerful and useful platform for a user managing numerous IoT devices on a single platform.

The whole system architecture is broken into four parts:

- **Hardware Layer**: All the IoT sensors and devices fall under this layer. Their primary task is data collection and sharing.

- **Network Layer**: The data that's collected by the sensors must reach the IoT cloud platform in some way and that is through this layer. This can be a direct transfer – that is, through Wi-Fi or Ethernet to the platform – or an indirect transfer, through an IoT gateway. Here, the hardware shares its data over a local network to the gateway (which can be a Raspberry Pi as well!) through BLE, WiFi, Zigbee, and so on, and then the data is shared to the IoT platform over the internet.

- **Middleware Layer**: The cloud platform's IoT services fall in this layer – that is, a fully managed IoT platform with numerous features and advantages over the conventional platforms.

- **Application Layer**: This is the final layer where all the applications fall and is what the end users use. They can be a customer, a data analyst – anyone. The application could be a mobile application, web dashboard, and so on.

Now, let's discuss two such platforms in brief, after which we will create a simple project, wherein we will demonstrate how to connect Node MCU to AWS IoT Core.

IoT cloud platforms

With the global IoT market rapidly progressing, an enterprise solution to handle all the IoT-related applications is essential. This is where the IoT cloud platforms come in. These platforms make the task of hosting an IoT service a lot easier as it only manages the hardware, software (in some use cases), and maintenance. All we need to do is host our service and write the product-specific code.

We will discuss two major IoT platforms in brief in this section: AWS and GCP.

Amazon Web Services

Amazon provides a range of services when it comes to IoT. The main advantage of these services is that they are completely managed and scalable and support almost an unlimited number of devices and message transfers. There are several use cases for the platform: industrial (IIoT), consumer and commercial markets, and more. Some popular IoT services that AWS provides are as follows:

- **AWS IoT Core**: This allows seamless and secure connections between devices over a fully managed architecture

- **AWS IoT Device Defender**: Security and auditing for your IoT devices

- **AWS IoT Device Management**: Register, monitor, and manage your IoT devices

- **FreeRTOS**: A lightweight operating system for microcontrollers for low-power edge devices

- **AWS IoT Greengrass**: For managing intelligent IoT devices on the edge

Google Cloud Platform

GCP allows the IoT Core service to link your IoT devices to the cloud. It is a highly scalable and fully managed service offered by GCP, which maintains a unique logical configuration for each device.

One of the defining features of this service is automatic load balancing, which means the architecture automatically scales up when the number of devices or the data traffic increases.

There are two major components of the IoT Core service:

- **Device Manager**: This manages all the connected IoT devices, including authentication, registration, and configuration management. It even helps you control your devices remotely from the cloud.

- **Protocol Bridge**: This component's main task is to provide a way to connect your IoT device to Google Cloud through various wireless communication protocols (for example, HTTP and MQTT).

Now that we've covered AWS and GCP, let's implement what we've discussed practically by connecting a Node MCU device to AWS IoT Core and sending BME280 sensor data to the cloud.

Project – getting started with AWS IoT Core

In this section, we will connect our Node MCU to a Wi-Fi network and, through that, connect it to AWS IoT Core. We will demonstrate the publish and subscribe capabilities by sending BME280 Sensor data to the cloud and sending a simple *Hello World* message from the cloud to the Node MCU development board.

Project hardware setup

To get started, we need to connect the BME280 sensor to the Node MCU. The following is a circuit diagram for the same:

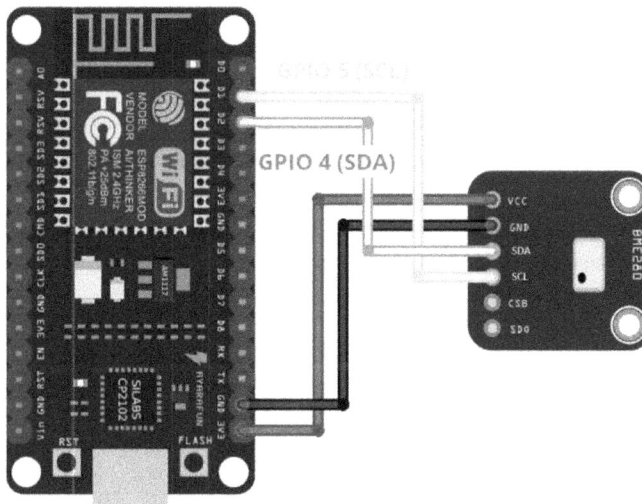

Figure 8.7 – NodeMCU BME280 schematic diagram

The preceding diagram shows that you need to connect the BME280 sensor to our NodeMCU development board. Now, let's set up the AWS IoT Core environment.

AWS IoT Core setup

First, we must set up AWS IoT Core. To do so, you must go to https://aws.amazon.com/:

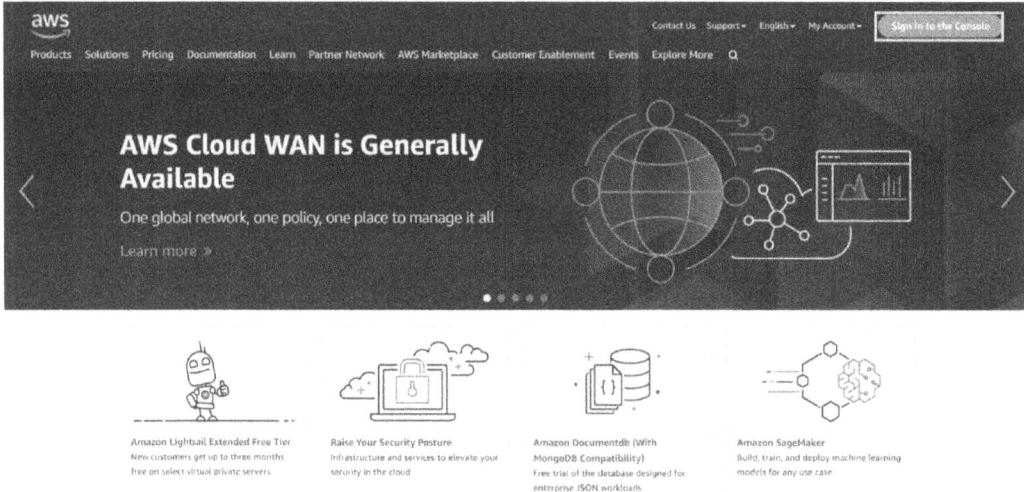

Figure 8.8 – AWS home page (click on the Sign in to the Console button)

Next, you must click on the **Sign in to the Console** button, which will redirect you to a new sign-in page. If you already have an active AWS account, just sign in using those user credentials; otherwise, just go to https://portal.aws.amazon.com/billing/signup, which will let you create a new account under the free tier.

Please note that you will need to enter your credit card details for the sign-up process, just for verification purposes. You will not be charged for anything until you use the services as you can stay within the free tier limits. Once you have signed in to your account, go to the AWS IoT Core service home page. You can do that either by searching for it via the **Console** search bar, going to the **Service** tab, or by going to https://aws.amazon.com/iot-core/.

This will open the AWS IoT dashboard, as shown in the following screenshot:

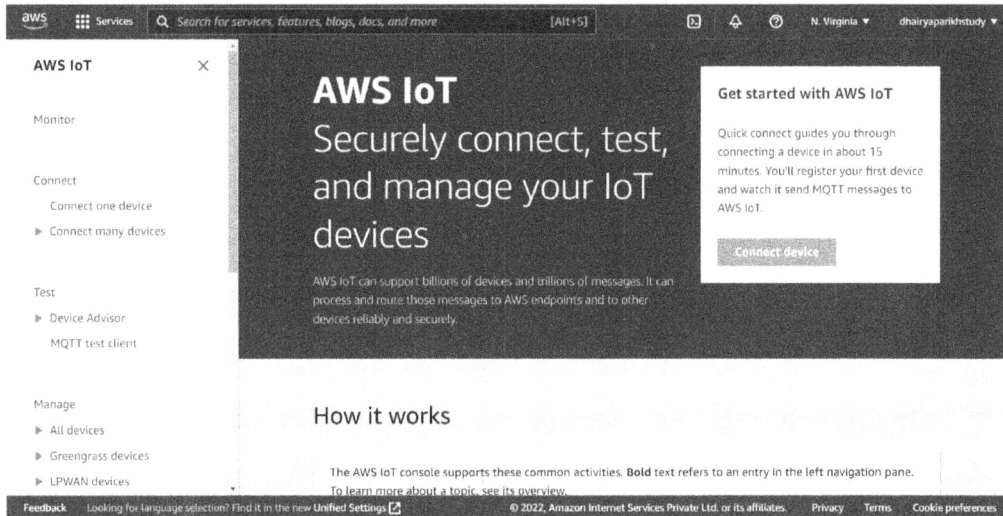

Figure 8.9 – The AWS IoT Core dashboard

Next, we have to connect our device to AWS IoT Core. For this, we need something called a **thing**, which is the representation of your physical device in the cloud. Just click on **All Devices** from the right pane and then click on **Things** from the dashboard home page. This will open a new web page that will take you through the thing registration process. Just click the **Create things** button:

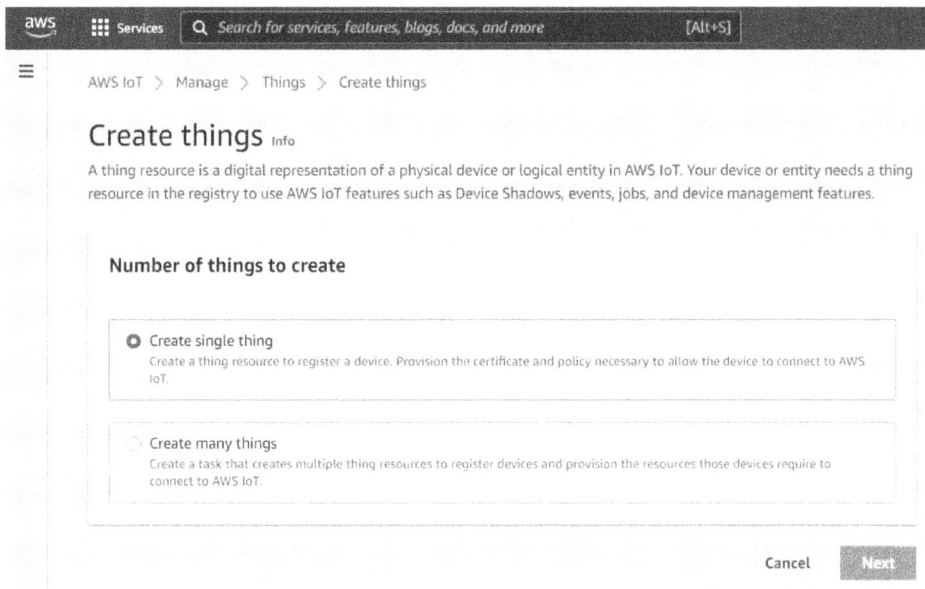

Figure 8.10 – Creating a new thing (AWS IoT)

We just need to create a single thing (for our Node MCU), so press **Next**. You will have to go through three steps to create a thing on AWS, as shown here:

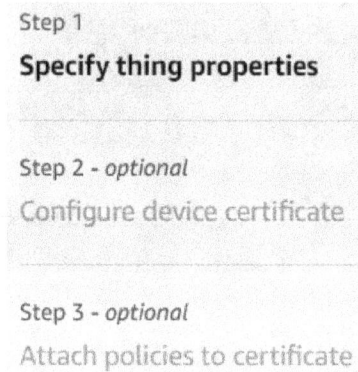

Figure 8.11 – Steps to create a thing on AWS IoT Core

In *Step 1*, all we need to do for now is give our thing a name. I will be naming it NodeMCU for this tutorial; you can name it whatever you like. There is no need to change any additional settings. Just press **Next** after you have given your thing an appropriate name:

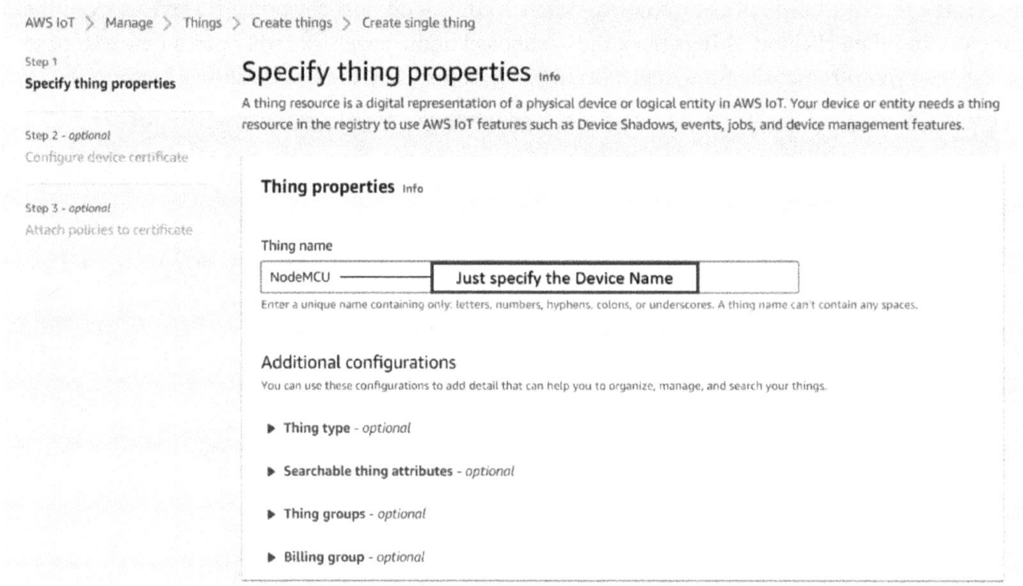

Figure 8.12 – Creating a new thing (Step 1)

The next step is to configure a certificate file (also known as a CSR) for your thing. This file is used for authentication purposes and is very important for maintaining security standards. We can either create a new certificate file (which we are going to do now) or upload an existing CSR file and skip this step.

We will just let AWS auto-generate the certificate for us (which is selected by default) and press **Next**:

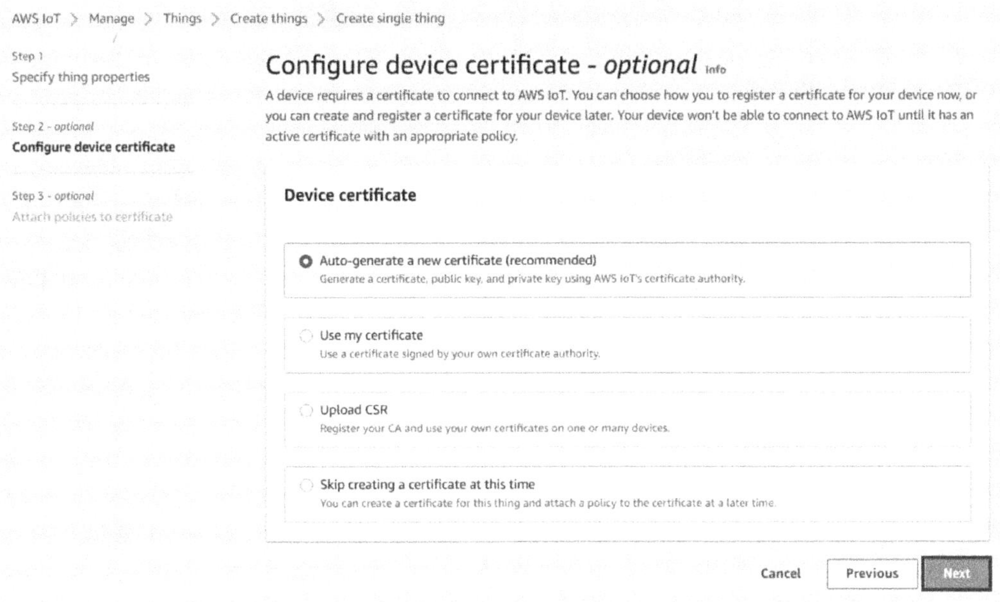

Figure 8.13 – Step 2 – Configure device certificate

The next step is to attach policies to this certificate, which means configuring what the device should be able to do (connect, subscribe, publish, and so on). For this tutorial, we need all three functionalities, so we will be creating our policy accordingly. Just click the **Create policy** button:

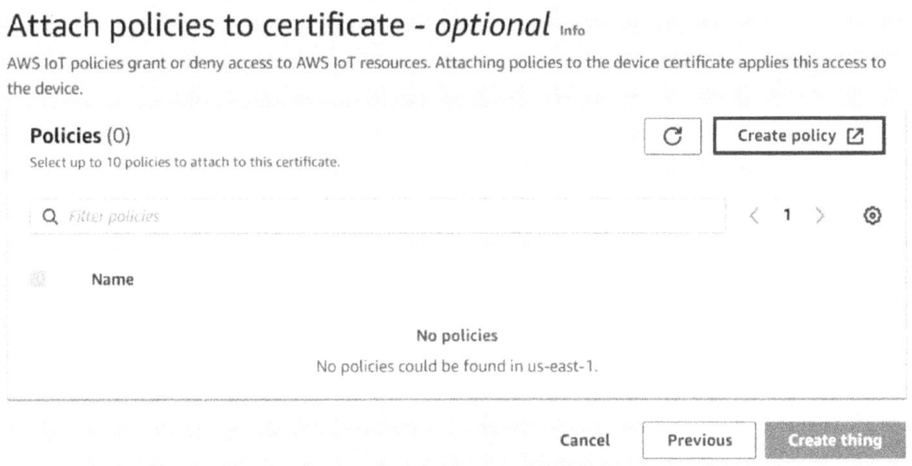

Figure 8.14 – Creating a new policy for our certificate

To create such a policy, we will need to add four policy actions. First, just give your policy a unique name (I have named mine `NodeMCU_Policy`), and then add the following actions:

- `Iot:Connect`
- `Iot:Publish`
- `Iot:Subscribe`
- `Iot:Receive`

Note that we need to allow all these policies. Moreover, we need the resource ARN for our thing for each policy. Getting this is a bit tricky, so the easy way is to choose a connect policy from the **Policy examples** tab; this will give you the resource ARN. Which looks something like this:

`arn:aws:iot:us-east-1:316634146583`

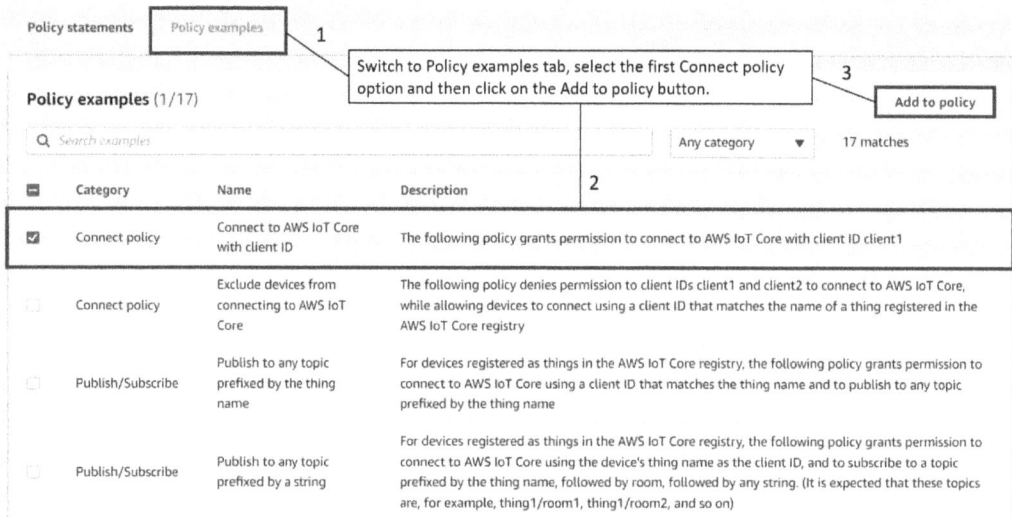

Figure 8.15 – Adding an example policy to get the resource ARN

Now, we need to fill in the following in the **Policy resource** section for each action:

- **Connect**: *
- **Publish**: *
- **Subscribe**: *
- **Receive**: *

Please note that if we want to add the Subscribe action, we need both the Receive and Subscribe actions. The first will allow the thing to receive messages, while the second will let you subscribe to any particular MQTT topic. * is a wildcard, which implies that any AWS resource that applies this policy will be able to access all the mentioned actions. Once all the actions have been set, the policy document will look something like this:

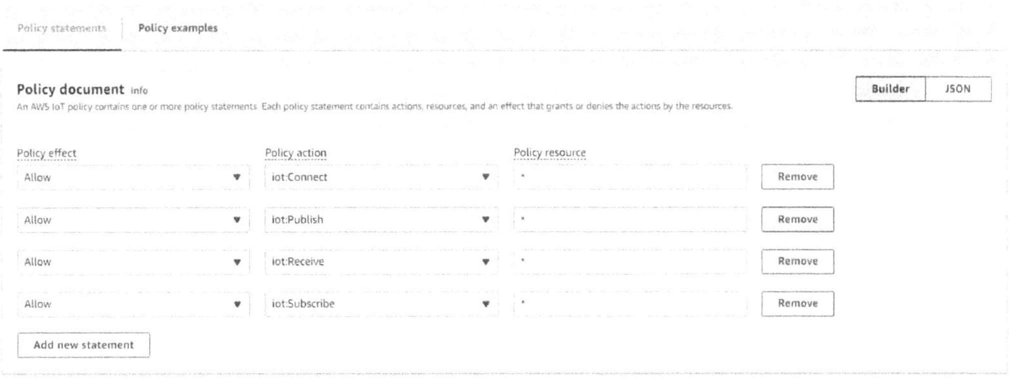

Figure 8.16 – Policy document configuration

After the configuration is complete, just click on the **Create policy** button. Your policy should appear in the **AWS IoT Policies** section, as well as the **Add policy to certificate** section (*Step 2*, remember?) for which we created this.

Just select the policy to attach it to the certificate and then press the **Create thing** button.

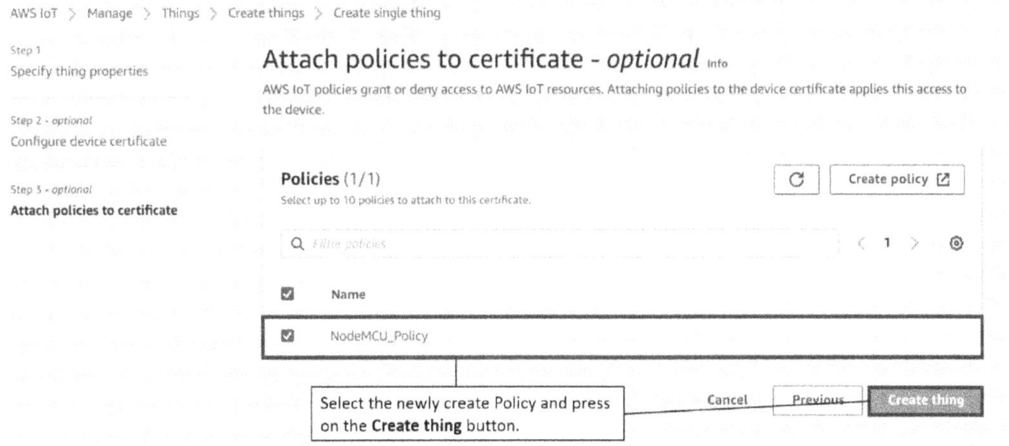

Figure 8.17 – Attaching our policy to the certificate and creating our thing

Now, a new window will pop up that lists all the key and certificate files that have been created. Here, we need to download four files:

1. Certificate file

2. Public key

3. Private key

4. The root CA 1 RSA 2,048-bit key

Please rename the files according to the names provided here to avoid confusion in the future. The following screenshot shows which files need to be downloaded for this tutorial:

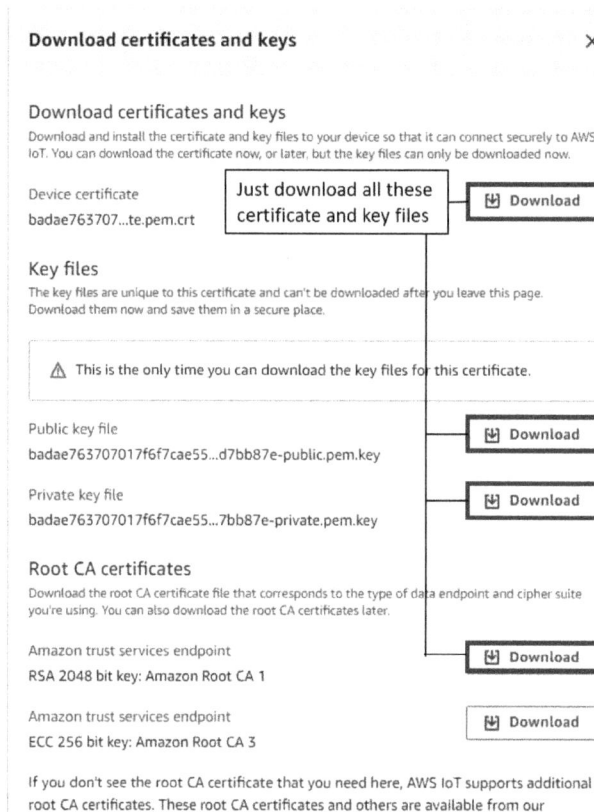

Download certificates and keys   ✕

Download certificates and keys

Download and install the certificate and key files to your device so that it can connect securely to AWS IoT. You can download the certificate now, or later, but the key files can only be downloaded now.

Device certificate   Just download all these   💾 Download
badae763707...te.pem.crt   certificate and key files

Key files

The key files are unique to this certificate and can't be downloaded after you leave this page. Download them now and save them in a secure place.

⚠ This is the only time you can download the key files for this certificate.

Public key file   💾 Download
badae763707017f6f7cae55...d7bb87e-public.pem.key

Private key file   💾 Download
badae763707017f6f7cae55...7bb87e-private.pem.key

Root CA certificates

Download the root CA certificate file that corresponds to the type of data endpoint and cipher suite you're using. You can also download the root CA certificates later.

Amazon trust services endpoint   💾 Download
RSA 2048 bit key: Amazon Root CA 1

Amazon trust services endpoint   💾 Download
ECC 256 bit key: Amazon Root CA 3

If you don't see the root CA certificate that you need here, AWS IoT supports additional root CA certificates. These root CA certificates and others are available from our

Figure 8.18 – The Download certificates and keys view

Once all the necessary files have been downloaded, just press the **Done** button. This concludes all the setup required from AWS IoT Core. Now comes the interesting part! We will write the code that will let the Node MCU board establish a connection with AWS IoT Core and allow us to publish and subscribe to the topics we mentioned during the setup.

For better reference, here are the publish and subscribe topic names we will be using for this project:

- **Publish Topic**: ESP8266/publish
- **Subscribe Topic**: ESP8266/subscribe

Let's start by explaining the code for this project.

Code explanation

The NodeMCU code we will be writing for this project will do the following:

1. Connect to a Wi-Fi network.
2. Establish a connection with AWS IoT Core.
3. Get the sensor data, create a JSON file, and publish it on the Publish topic.
4. Subscribe and print any messages that have been published on the Subscribe topic.

First, we will do something different in this project, which is a better practice than what we have done so far. We will store all the sensitive data information such as the certificates, private key, Wi-Fi credentials, and the AWS endpoint address in a separate `secrets.h` file for more secure and flexible usage of these parameters in our main code.

To do this, just create a new tab after creating a new Arduino file by going to the downward arrow icon at the top-right of the Arduino IDE window and expanding the menu to find and click the **Create New Tab** option. Save the file with the name `secrets.h`.

Now, let's look at a step-by-step explanation of the code, as we have done in all the previous projects. We have two files, so we will divide this section into two parts as well. Let's start with the `secrets.h` file:

```
#include <pgmspace.h>
#define SECRET
const char WIFI_SSID[] = "…";              //Wifi SSID Name
const char WIFI_PASSWORD[] = "….";          //Wifi Password
#define THINGNAME "…." // Name of your thing in AWS
int8_t TIME_ZONE = +5.5; //IST(India): +5.5 UTC (Time Zone
parameter)
const char MQTT_HOST[] = "xxxxxxxxxxxxxx-ats.iot.us-east-1.
amazonaws.com"; // Get this from the AWS IoT Core Settings
static const char cacert[] PROGMEM = R"EOF(
-----BEGIN CERTIFICATE-----
[…………]
-----END CERTIFICATE-----
)EOF";
// Copy contents from XXXXXXX-certificate.pem.crt here ▼
```

```
static const char client_cert[] PROGMEM = R"KEY(
-----BEGIN CERTIFICATE-----
[............]
-----END CERTIFICATE-----

)KEY";
// Copy contents from  XXXXXXX-private.pem.key here ▼
static const char privkey[] PROGMEM = R"KEY(
-----BEGIN RSA PRIVATE KEY-----
[............]
-----END RSA PRIVATE KEY-----

)KEY";
```

This file is used to store all the sensitive and reusable parameters of the main code. Let's look at these parameters:

1. First are the Wi-Fi credentials, which must be specified.

2. Next, you need to store the name of the thing you gave on AWS IoT as this will be used as the client's name to establish a connection.

3. The TIME_ZONE parameter (the difference between your time zone and UTC in hours) will be used to fetch the latest UTC (covered later in the code).

4. MQTT_HOST will contain the AWS endpoint, which will be used as the MQTT host. It can be fetched by going to the settings of AWS IoT Core. The following screenshot shows where you can find it:

Figure 8.19 – Getting the AWS endpoint for the project

5. Finally, we need to store the certificate, private key, and the Amazon Root CA1 certificate we downloaded from AWS. To get the text data of these certificates, just open them in Notepad and copy and paste their contents into this file accordingly. You will find comments to help you out with this.

Next, we will write the main code for this project, in which we will include this `secrets.h` file so that we have access to all the variables and constants we defined in it.

Chapter_8_Project_Code.ino

Since this code is quite large, we have broken it into short snippets for easier understanding and explanation. First, we must import the necessary libraries:

```
#include <ESP8266WiFi.h>
#include <WiFiClientSecure.h>
#include <PubSubClient.h>
#include <ArduinoJson.h>
#include <time.h>
#include <Wire.h>
#include <Adafruit_BMP280.h>
#include "secrets.h"
```

As you can see, we will be using some new libraries we haven't covered:

- The `WiFiClientSecure.h` file is a subpart of the ESP8266WiFi library and will be used to connect to AWS IoT Core using the certificate and private key we downloaded. To learn more about this, go to `https://arduino-esp8266.readthedocs.io/en/latest/esp8266wifi/bearssl-client-secure-class.html`.

- The **ArduinoJSON** library will be used to format the sensor data in JSON format before it's published on our topic. This is a good practice and can be implemented in the projects we covered previously in this book as well. Please note that you will have to download ArduinoJSON version 6.0 or higher for this project. Please visit the library documentation to learn more about this library: `https://arduinojson.org/v6/doc/`.

- The `time.h` file is used to get access to certain functions to set the time as per your time zone for the NodeMCU board. This is important as the current time is required for certification verification of AWS IoT.

Now, let's look at the variable, object, and constant declarations:

```
// BMP 280 Sensor value variables
float temperature_C ;
float pressure ;
```

```
float altitude ;
// Variables to implement publish every 5 second logic
unsigned long lastMillis = 0;
unsigned long previousMillis = 0;
const long interval = 5000;
// variables to store the latest time fetched by the NTP
Client.
time_t now;
time_t nowish = 1510592825;
// AWS Publish and subscribe topic mentioned in the AWS policy
you created
#define AWS_IOT_PUBLISH_TOPIC    "ESP8266/publish"
#define AWS_IOT_SUBSCRIBE_TOPIC "ESP8266/subscribe"
// WiFi SSL to add the certificates and key we retrieved from
AWS
WiFiClientSecure net;
BearSSL::X509List cert(cacert);
BearSSL::X509List client_crt(client_cert);
BearSSL::PrivateKey key(privkey);
// Objects for pubsub and BMP sensor
PubSubClient client(net);
Adafruit_BMP280 bmp;
```

The next step is to initialize all the necessary variables, constants, and objects that we will be using in this code. Certain variables have been used without any initializations (their style has been changed to bold). This is because they have been defined in the secrets.h file, which we will define at the end.

Now, let's look at the NTPConnect function:

```
void NTPConnect(void)
{
  Serial.print("Setting time using SNTP");
  configTime(TIME_ZONE * 3600, 0 * 3600, "pool.ntp.org", "time.
nist.gov");
  now = time(nullptr);
  while (now < nowish)
  {
    delay(500);
    Serial.print(".");
```

```
    now = time(nullptr);
  }
  Serial.println("done!");
  struct tm timeinfo;
  gmtime_r(&now, &timeinfo);
  Serial.print("Current time: ");
  Serial.print(asctime(&timeinfo));
}
```

This function is used to fetch the latest time. This is required for proper authentication, as discussed previously. Please note that the only thing you need to keep in mind here is the TIME_ZONE variable, which will also be defined in the secrets.h file. This is the number by which the time in your region is behind or ahead of the UTC (as we need UTC for authentication).

Now, let's look at the MQTT callback function:

```
void messageReceived(char *topic, byte *payload, unsigned int
length)
{
  Serial.print("Received [");
  Serial.print(topic);
  Serial.print("]: ");
  for (int i = 0; i < length; i++)
  {
    Serial.print((char)payload[i]);
  }
  Serial.println();
}
```

This is the standard MQTT callback function we have been using throughout this book. It prints the incoming messages on the subscribed topics on the Serial monitor.

Now, let's look at the Setup_WiFi() function:

```
void Setup_WiFi()
{
  delay(10);
  // We start by connecting to a WiFi network
  Serial.println(String("Attempting to connect to SSID: ") +
String(WIFI_SSID));
```

```
    WiFi.mode(WIFI_STA);
    WiFi.begin(WIFI_SSID, WIFI_PASSWORD);
    while (WiFi.status() != WL_CONNECTED)
    {
      delay(500);
      Serial.print(".");
    }
    randomSeed(micros());
    Serial.println("");
    Serial.println("WiFi connected!");
    Serial.println("IP address: ");
    Serial.println(WiFi.localIP());
}
```

The main purpose of this function is to establish a network connection over Wi-Fi using the credentials you will specify in the `secrets.h` file (`WIFI_SSID` and `WIFI_PASSWORD`).

Now, let's look at the `connectAWS()` function:

```
void connectAWS()
{
  delay(3000);
  Setup_WiFi();
  NTPConnect();
  net.setTrustAnchors(&cert);
  net.setClientRSACert(&client_crt, &key);
  client.setServer(MQTT_HOST, 8883);
  client.setCallback(messageReceived);
  Serial.println("Connecting to AWS IOT");
  while (!client.connect(THINGNAME)) {
    Serial.print(".");
    delay(1000);
  }
  if (!client.connected()) {
    Serial.println("AWS IoT Timeout!");
    return;
  }
  // Subscribe to a topic
```

```
  client.subscribe(AWS_IOT_SUBSCRIBE_TOPIC);
  Serial.println("AWS IoT Connected!");
}
```

This function does the following:

1. Connects to Wi-Fi by calling the `setup_wifi` function.

2. Gets the latest time by calling the `NTPConnect` function.

3. Establishes a connection with the AWS MQTT endpoint (this will be fetched from the AWS IoT Core settings) and initializes the callback function.

4. Tries to establish a connection with the thing we created. If a connection can be established, it subscribes to the topic we have specified in our AWS IoT policy (the topic name is already defined).

Now, let's look at the `publishMessage()` function:

```
void publishMessage()
{
  StaticJsonDocument<200> doc;
  doc["time"] = millis();
  doc["temperature"] = temperature_C;
  doc["pressure"] = pressure;
  doc["altitude"] = altitude;
  char jsonBuffer[512];
  serializeJson(doc, jsonBuffer); // print to client
  client.publish(AWS_IOT_PUBLISH_TOPIC, jsonBuffer);
}
```

This function's primary task is to store the latest BMP sensor values in a JSON document, convert it into a `char` array, and then publish this data on the AWS Publish topic we defined in the code.

Now, let's look at the `setup()` function:

```
void setup()
{
  Serial.begin(115200);
  if (!bmp.begin(0x76)) {
    Serial.println(F("Could not find a valid BMP280 sensor,
check wiring or "
                     "try a different address!"));
```

```
   while (1) delay(10);
}
// Default settings from datasheet
bmp.setSampling(Adafruit_BMP280::MODE_NORMAL,
               Adafruit_BMP280::SAMPLING_X2,
               Adafruit_BMP280::SAMPLING_X16,
               Adafruit_BMP280::FILTER_X16,
               Adafruit_BMP280::STANDBY_MS_500);
connectAWS();
}
```

Here, we do the following:

1. Initialize or open a Serial connection with a baud rate of 115200.
2. Check if the BMP sensor is connected, and if it is, specify the default setting parameters for it.
3. Call the connectAWS function, which connects to Wi-Fi, sets up the latest time, and connects to AWS IoT Core.

Now, let's look at the loop() function:

```
void loop()
{
  temperature_C = bmp.readTemperature();
  pressure = bmp.readPressure();
  altitude = bmp.readAltitude(1013.25);
  Serial.print(F("Temperature : "));
  Serial.print(temperature_C);
  Serial.print(F("%  Pressure : "));
  Serial.print(pressure);
  Serial.print(F("   Altitude : "));
  Serial.println(altitude);
  delay(2000);
  now = time(nullptr);
  if (!client.connected())
  {
    connectAWS();
  }
  else
```

```
  {
    client.loop();
    if (millis() - lastMillis > 5000)
    {
      lastMillis = millis();
      publishMessage();
    }
  }
}
```

This is the final `loop()` function in which we will fetch the latest BMP sensor values and print them on the Serial monitor and set up the latest time. After that, we will check whether a connection with AWS IoT Core has been established and, if it has, publish the sensor values on the AWS Publish topic (ESP8266/publish in this case) every 5 seconds.

With that, we've explained the code. Now, we will see the project in action, for which we will be using the MQTT test client provided by AWS IoT.

Project demonstration

Now that everything has been set up, we are ready to test the project! Once you have connected the BMP280 sensor to the NodeMCU board, just upload the code to the board. If everything has been done according to this tutorial, you will be able to upload the code onto the board successfully. This completes the setup on the hardware end.

Next, you will have to open the AWS IoT Core dashboard from the left pane and select the **MQTT test client** option, which you can find under the **Test** section. This will open a new page that will look something like this:

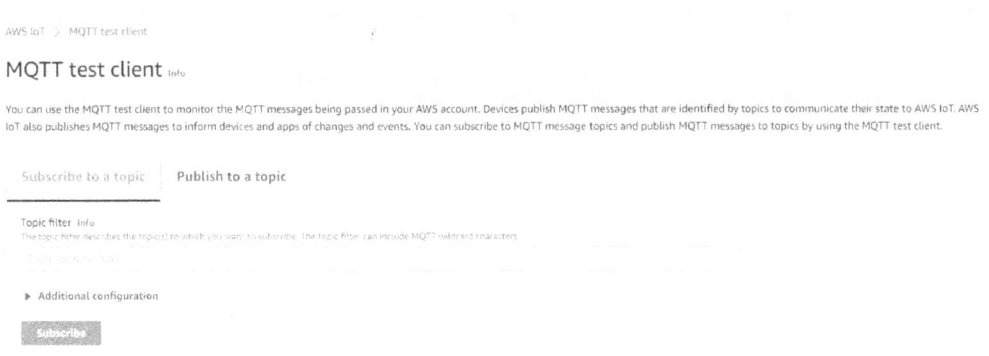

Figure 8.20 – The MQTT test client page

This is a web-hosted MQTT client that has already been connected to your AWS endpoint. You can publish and subscribe to different topics through this tool. Now, let's check whether we can receive the BMP280 sensor values from our NodeMCU. For that, just add the **ESP8266/publish** text under **Topic filter** in the **Subscribe to a topic** section and click the **Subscribe** button.

Now, connect the NodeMCU to a power source and wait for a few seconds (setting up the latest time takes some time). You should be able to see the sensor values on the screen every 5 seconds:

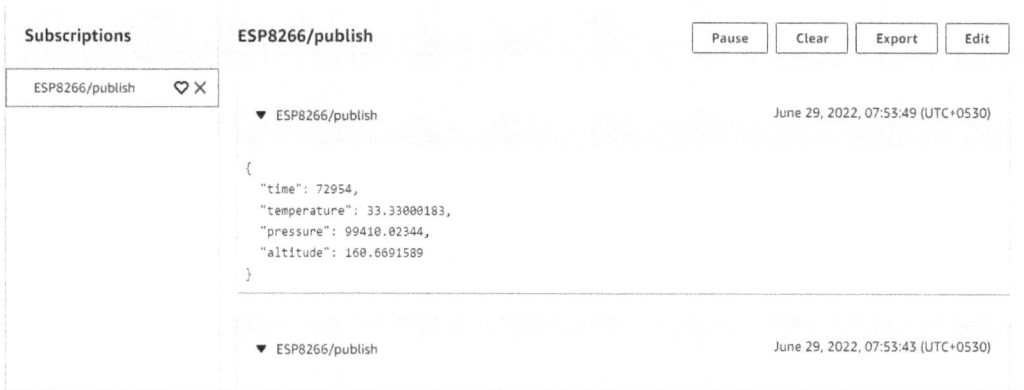

| Subscriptions | ESP8266/publish | | Pause | Clear | Export | Edit |

ESP8266/publish ♡ ✕

▼ ESP8266/publish   June 29, 2022, 07:53:49 (UTC+0530)

```
{
  "time": 72954,
  "temperature": 33.33000183,
  "pressure": 99410.02344,
  "altitude": 160.6691589
}
```

▼ ESP8266/publish   June 29, 2022, 07:53:43 (UTC+0530)

Figure 8.21 – NodeMCU sensor values every 5 seconds

Great! Now, let's test whether our NodeMCU board can print messages that have been published on the **ESP8266/subscribe** topic (specified in the code). For this, just open the **Publish to a topic** section on the MQTT test client tool and type the aforementioned topic name in the **Topic name** textbox. Keep the message payload as is. Before publishing the message, open the Serial monitor for your NodeMCU board as you will be able to see the message there.

Once you have it open, click the **Publish** button. You should be able to see the message there, as shown here:

```
Temperature : 33.68%   Pressure : 99475.84   Altitude : 155.11
Temperature : 33.68%   Pressure : 99475.66   Altitude : 155.12
Temperature : 33.67%   Pressure : 99474.95   Altitude : 155.18
Received [ESP8266/subscribe]: {
  "message": "Hello from AWS IoT console"
}
Temperature : 33.67%   Pressure : 99474.60   Altitude : 155.21
Temperature : 33.67%   Pressure : 99474.25   Altitude : 155.24
Temperature : 33.68%   Pressure : 99474.44   Altitude : 155.22
```

Figure 8.22 – Sample message published through the MQTT test client on AWS

This marks the end of this section. The possibilities with this are endless as we can pass this data through any of the available AWS services to create dashboards, run inference by passing them through trained ML models, and even use this data to train a machine learning model. We can even create automations using AWS Lambda – this is just the tip of the iceberg. This is a step you will take only after you have scaled your projects to a certain level as these enterprise solutions are pretty expensive and the complexity of the solutions also increases after adding different AWS services into the mix. Now, let's learn how to scale them immediately and on a much smaller scale.

How to scale your projects using the current hardware

We discussed a lot of options for scaling our current setup, starting with using a cloud-hosted MQTT broker, in *Chapter 7, Taking Your MQTT Broker Global*. In the previous section, we discussed how we can use enterprise IoT platforms for this. But the main drawback of such products is that they are quite expensive and although they do provide customizations, there are always some drawbacks you may face while working on a particular use case.

For this reason, I have added this section as well. This will give you a gentle introduction to all the tools you will need to build an actual product using the existing tools that you have. For maximum customization, you always need to have your own software base, which you will have to develop from scratch. There are several approaches that you can take to achieve this, and, in this section, we will cover a few of them.

So, without further ado, let's get started.

Home Assistant

Node-RED, though easy to use, has its limitations as it was created for educational purposes. Moreover, there are far more capable software packages built specifically for home automation such as Home Assistant, openHAB, and IFTTT.

In this section, we will discuss one of the most popular and obvious choices – **Home Assistant**:

> *"Home Assistant is a dedicated home automation system that is completely open source. It allows you to control all your devices over a local network (which provides data privacy) by running it on a local server such as a Raspberry Pi. The major advantage of this software is the vast developer community that supports this project."*

This system provides several advantages over the existing Node-RED system that we use:

- Its major advantage is the number of supported devices, specifically commercial home automation devices such as Philips Hue smart lights, Google's Nest thermostat, and more. All such devices can be configured with our current setup without any hassle. All such connections are referred to as Configurations in Home Assistant. At the time of writing, there are 1,800+ such configurations available and they keep on adding more.

- We can even add different wireless connectivity modes to our system just by connecting an external board to our Raspberry Pi. For instance, you can add Zigbee support to the system just by adding a Zigbee support board to your Raspberry Pi.

- One of the most impactful advantages of using Home Assistant is the developer community. It has a very active and helpful community that can support you every step of the way.

Now, to get started with Home Assistant on the Raspberry Pi, you just need to create an OS image to specifically run the Home Assistant application. To do this, you must follow some steps that are similar to those you followed to set up your SD card in *Chapter 1, Introduction to Raspberry Pi and MQTT* (*Setting up the SD card*) with a small change:

1. When you open the Raspberry Pi Imager and click on the **Choose OS** button, you have to select the **Home Assistant** option instead of the latest Raspberry Pi OS build. The following screenshot shows how to select it:

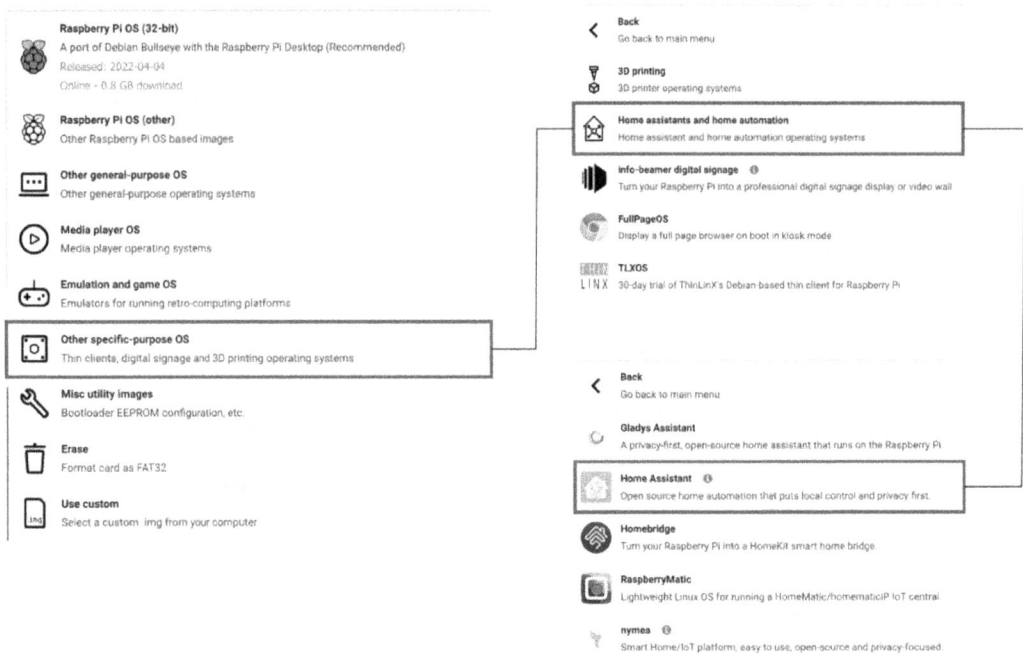

Figure 8.23 – Installing and flashing the Home Assistant OS on the Raspberry Pi

2. Once you select that option, you will have two options for operating systems – one for Raspberry Pi 3 and the other for Raspberry Pi 4/400. Just choose one according to the Raspberry Pi model you own and then follow the rest of the steps, as you did in *Chapter 1, Introduction to Raspberry Pi and MQTT*. The following screenshot shows these options:

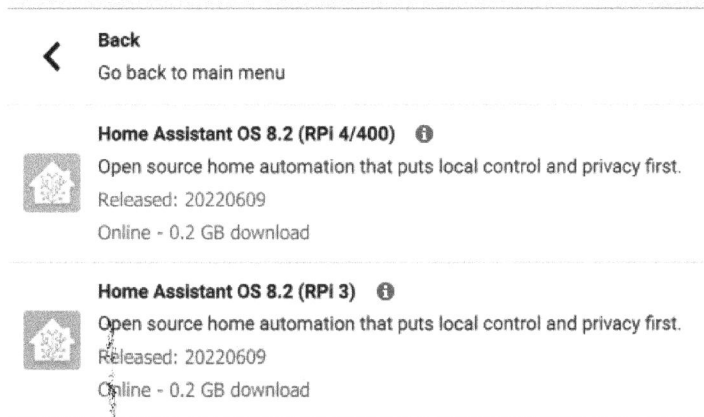

Figure 8.24. – Home Assistant OS options based on the Raspberry Pi model

Now that you have the Home Assistant OS flashed on your SD card, just insert that into the Raspberry Pi so you can get started with Home Assistant! Here is the link to the official Home Assistant website, which contains several examples, detailed documentation, and more to help you get started: `https://www.home-assistant.io/`.

Now, if you still want further customization, you will have to install and set up every software requirement yourself from scratch. We will explore this option in brief and learn how to set up a basic LAMP server on the Raspberry Pi, which is the first step you need to take toward developing a fully custom setup.

So, let's get started!

LAMP Server

LAMP stands for **Linux Apache MySQL PHP/Python/Pearl**. This is a popular software bundle used specifically for web development. There is a very popular alternative for the programming language, JavaScript, a very popular and powerful web development language.

We will stick to the basics and set up the PHP language, along with an additional piece of software, phpMyAdmin, which is a database management web interface. For the database, we will use MariaDB, which is built on top of MySQL with some additional and useful features. Follow these steps:

1. **Updating the Raspberry Pi OS**: Before we can install any component, we need to make sure that our OS (the Linux component) is up to date. Just run the `update` and `upgrade` commands that we have run several times throughout this book in a new terminal window:

```
sudo apt update && sudo apt upgrade -y
```

2. **Installing Apache**: Once this is done, we will start by installing and setting up our first component, Apache. It is a popular web server software that allows you to host and handle web pages. To install this on your Raspberry Pi, just run the following command:

```
sudo apt install apache2 -y
```

The following screenshot shows what the output of this command looks like:

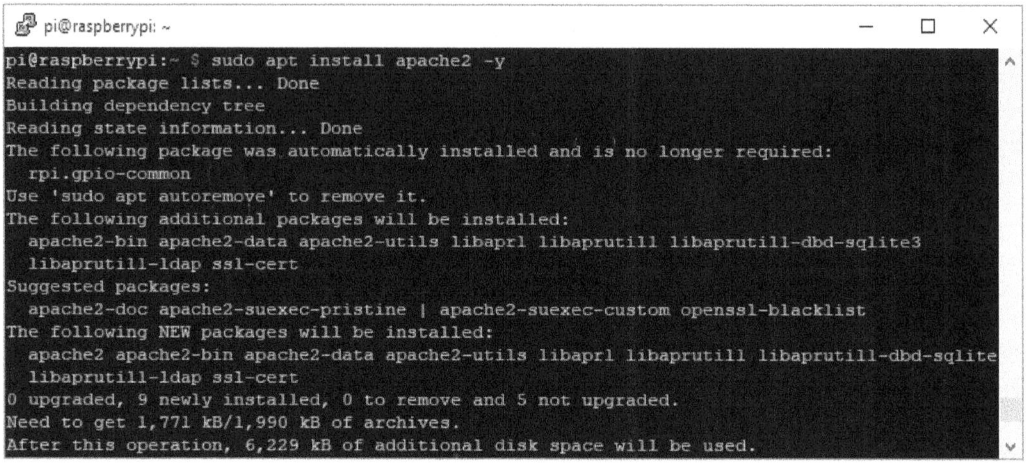

Figure 8.25 – Installing Raspberry Pi Apache2

Once this command runs successfully, Apache will be installed. Now, to test this installation, you just need to do the following:

1. First, change the directory to /var/www/html. Just type the following command in the terminal:

```
cd /var/www/html
```

2. This is the home directory of your server. All the source code for your website or web page will go in here. By default, an index.html file is present in this directory. You can easily access your server's index file through your Raspberry Pi's IP address. You can easily get the address by typing ifconfig or hostname -L in your terminal.

3. Once you get the IP address, just type that address into any browser connected to the same network as your Pi. You should see the following output if everything has been installed correctly:

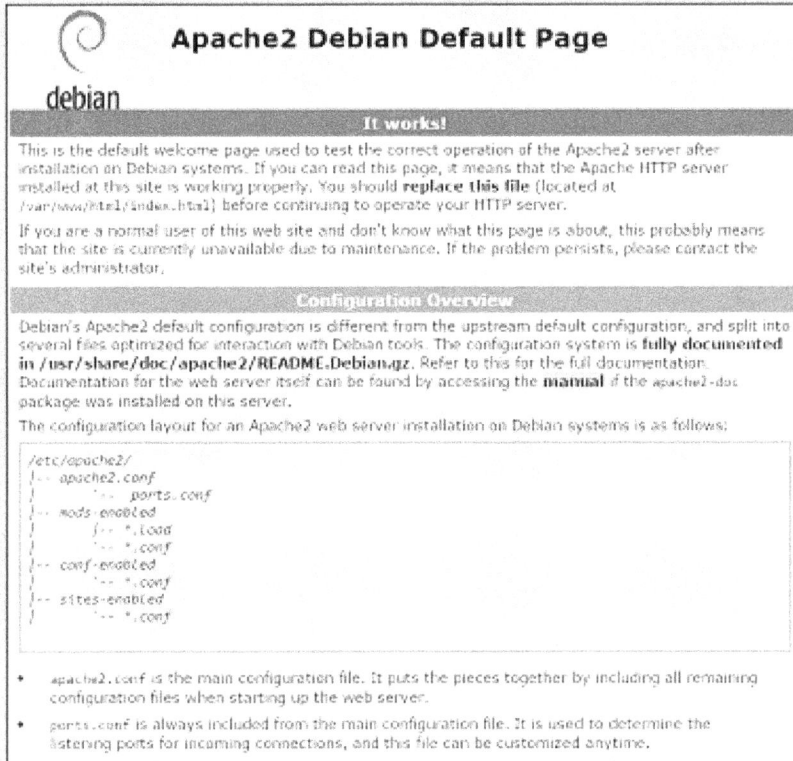

Figure 8.26 — Apache2 Debian Default Page (index.html)

With that, we have installed the Apache package. Next, we will install the PHP component.

PHP installation and setup

Hypertext Preprocessor (**PHP**) is a server-side scripting language that is used to develop dynamic web applications. To install PHP on your Pi, just type the following command in your terminal:

```
sudo apt install php -y
```

Now, let's test whether it has been installed as expected. For this, we will replace the default index. html file with a file of our own. For this, just move to Apache's home directory and type in the following commands:

```
cd /var/www/html
sudo rm index.html
sudo nano index.php
```

In your `index.php` file, add the following code to print a message of your choice on your web page. We will print the most basic message – `"hello world"`:

```
<?php echo "hello world"; ?>
```

The following screenshot shows what this will look like on your screen:

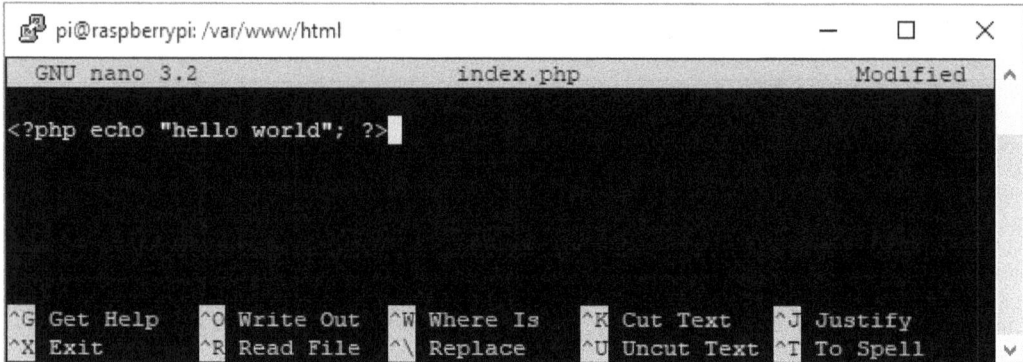

Figure 8.27 – Your first PHP script!

To save your file, press *Ctrl + X*, followed by *Y*, and press *Enter* to exit. Now, for the changes to take effect, we will restart the Apache server using the following command:

```
sudo service apache2 restart
```

To test this, just type in the IP address of your Pi on a browser. You should see an output similar to the following instead of the one you saw when you first installed Apache:

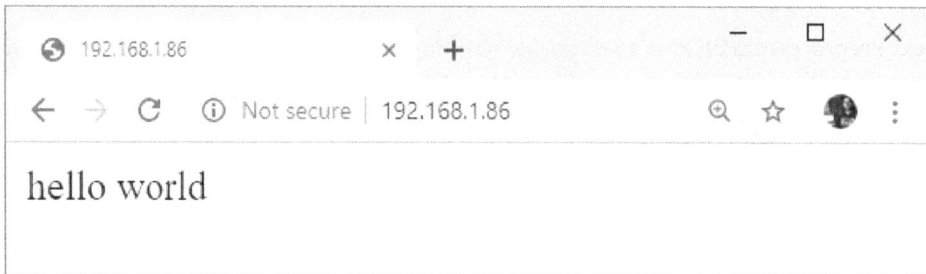

Figure 8.28 – Sample output page

If everything works as expected, then you have successfully installed and configured PHP on your Pi. Now, you can write PHP scripts that can be used to communicate with databases and client-side scripting language pages.

In the next subsection, you will learn how to install MySQL on your Raspberry Pi. Essentially, you will be installing a database server on your Pi that can be accessed through any device connected to the local network. Moreover, you will install another software component called phpMyAdmin, which is a GUI for accessing and managing the database.

MariaDB and phpMyAdmin installation

MySQL (**SQL** stands for **Structured Query Language**) is a popular relational database that is completely open source.

We will be installing the MariaDB server, which has been built on top of MySQL. First, we need to install MariaDB and the `php-mysql` package, which will allow us to use phpMyAdmin to manage our MariaDB database. Now, to install and set up everything, we need to run three commands, one after the other:

```
sudo apt install mariadb-server php-mysql -y
sudo service apache2 restart
sudo mysql_secure_installation
```

The first command will install the database server and the supporting package. Then, we need to restart the Apache web server so that it can detect the new packages. To complete the installation, we need to type in the third command, which will define the configuration:

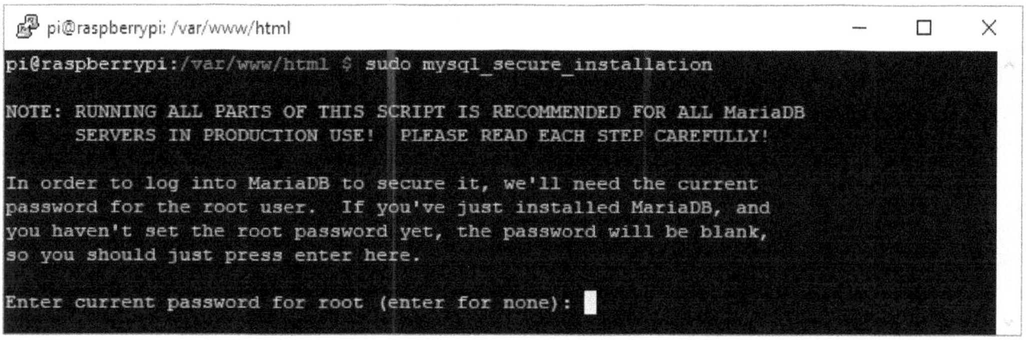

```
pi@raspberrypi: /var/www/html                                    —    □    ✕

pi@raspberrypi:/var/www/html $ sudo mysql_secure_installation

NOTE: RUNNING ALL PARTS OF THIS SCRIPT IS RECOMMENDED FOR ALL MariaDB
      SERVERS IN PRODUCTION USE!  PLEASE READ EACH STEP CAREFULLY!

In order to log into MariaDB to secure it, we'll need the current
password for the root user.  If you've just installed MariaDB, and
you haven't set the root password yet, the password will be blank,
so you should just press enter here.

Enter current password for root (enter for none): █
```

Figure 8.29 – The mysql_secure_installation command's output

As you can see, this command lets you secure your database using the credentials that you specify. Just follow these steps to complete this process:

1. You will be asked to **Enter current password for root**. There's nothing to add here, so just press *Enter*.

2. Type in *Y* and press *Enter* to set the root password.

3. Type in a password at the new password prompt and press *Enter*.

> **Important Note**
>
> Remember this root password as you will need it later.

4. Type *Y* to remove anonymous users.

5. Type *Y* to disallow root login remotely.

6. Type *Y* to remove the test database and access to it.

7. Type *Y* to reload the privilege tables.

When the installation is completed, you'll see a message stating, `Thanks for using MariaDB!`. This can be seen in the following screenshot:

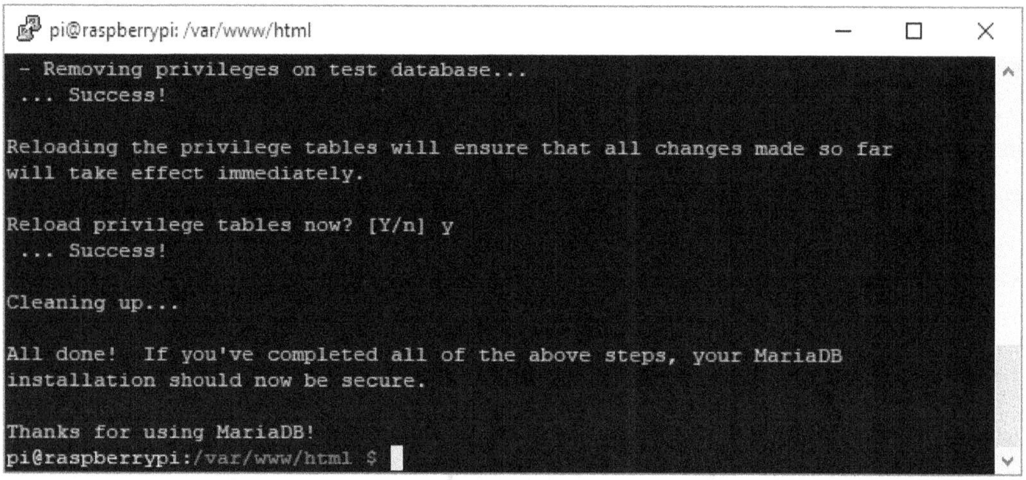

Figure 8.30 – MariaDB installation completion message

Next, we will install `phpmyadmin` on our system. To do that, just type the following command in your terminal:

```
sudo apt install phpmyadmin -y
```

The PhpMyAdmin installation program will ask you a few questions. We'll use `dbconfig-common` to configure this. Then, follow these steps:

1. Select **Apache2** when prompted and press the *Enter* key.

2. When asked **Configuring phpmyadmin?**, select **OK** and press *Enter*.

3. When asked **Configure database for phpmyadmin with dbconfig-common?**, select **Yes**.

4. Type your password and press **OK**:

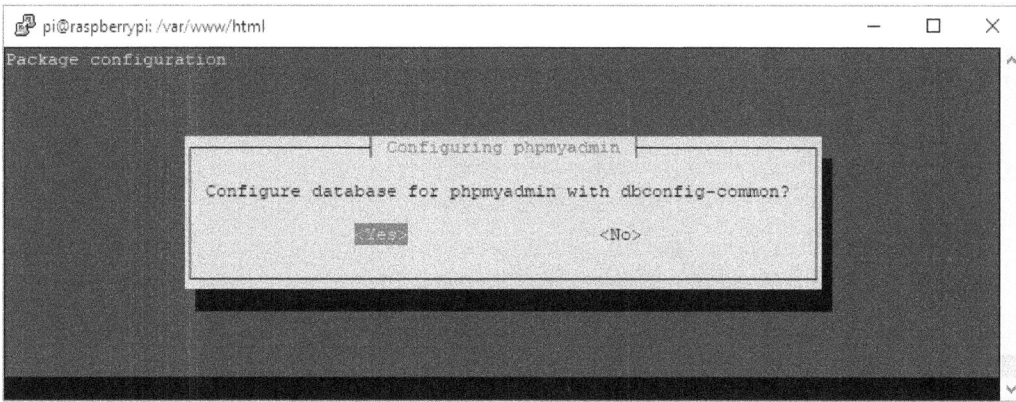

Figure 8.31 — phpMyAdmin configuration window

Now, we just need to enable the PHP MySQLi extension and restart Apache2 for changes to take effect. This can be done by using the following commands:

```
sudo phpenmod mysqli
sudo service apache2 restart
```

Now, we need to perform one final step before we can open the phpMyAdmin interface. We need to create a link that will allow Apache's home directory to access the phpmyadmin folder. This can be done by the following command:

```
sudo ln -s /usr/share/phpmyadmin /var/www/html/phpmyadmin
```

Now, you will be able to see a phpmyadmin folder in Apache's home directory. This means you can access the UI for this application just by typing the following address in any browser connected to the same network as your Raspberry Pi: <Pi's IP Address>/phpmyadmin.

You should see the login page for phpMyAdmin open in your browser window, as shown in the following screenshot:

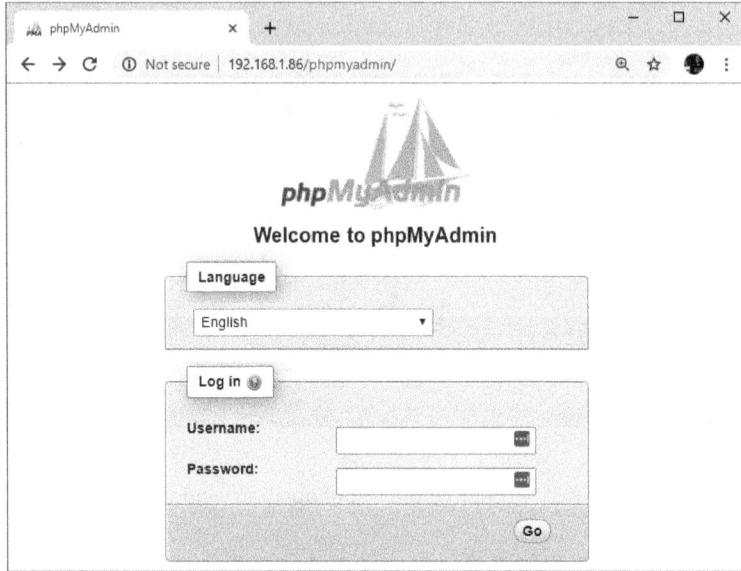

Figure 8.32 – phpMyAdmin login page

Enter the defined user credentials and press the **Go** button to log in:

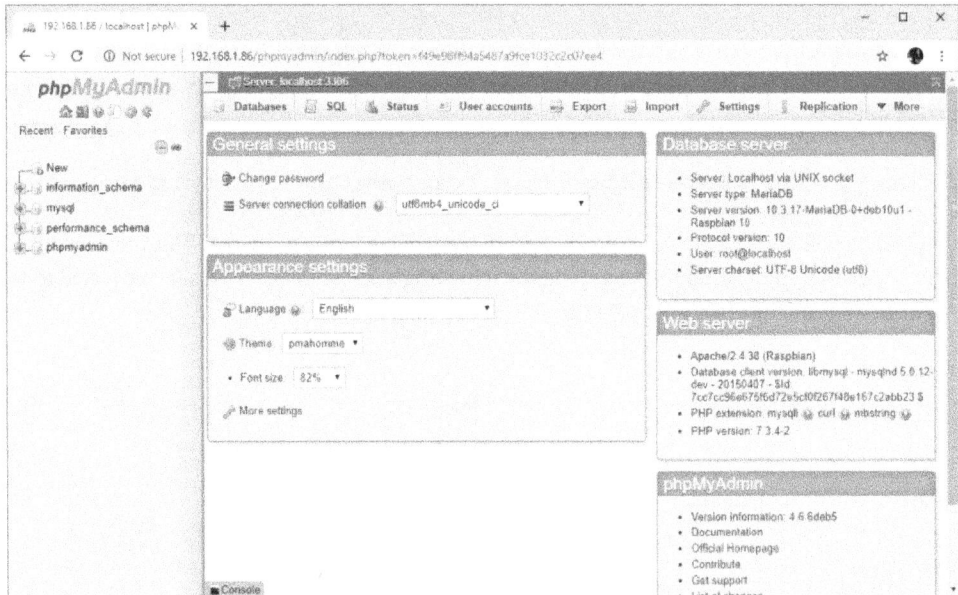

Figure 8.33 — PhpMyAdmin home page

With that, you have installed both the database and the database management component and, in turn, the LAMP configuration on our Raspberry Pi!

Now, you can use these software tools to create dashboards and access them through a web interface. This gives you immense customization options, but the tradeoff is that you need to know how to create web pages in any programming language of your choice.

"All good things must come to an end."

Hence, this marks the end of our book. Now, let's summarize what we have learned in this book.

Summary

We covered a lot of topics throughout this book and provided several projects to help you practically implement the knowledge you've gained.

You have learned a variety of essential skills throughout this book. For instance, you can now set up your own Raspberry Pi and you know what MQTT is, which means you can use this communication protocol in any of your projects. Furthermore, you have learned how to set up the Node MCU and ESP32 development boards and how to write efficient and robust code for them. At this point, you can build two fairly complex prototype projects on your own: an IoT Weather Station and a Smart Home control system!

After that, you learned the basics of Bash (the language used to write and execute Linux commands) and how to set up an online MQTT broker, either on an independent provider or on your very own virtual machine. Finally, you learned how to connect your IoT devices to AWS IoT and how to connect your LAMP server to your Raspberry Pi.

These are just some of the things you can do now. But this is just the beginning – keep learning and improving yourself!

Index

U

V

W

X

‹packt›

Packt.com

Subscribe to our online digital library for full access to over 7,000 books and videos, as well as industry leading tools to help you plan your personal development and advance your career. For more information, please visit our website.

Why subscribe?

- Spend less time learning and more time coding with practical eBooks and Videos from over 4,000 industry professionals

- Improve your learning with Skill Plans built especially for you

- Get a free eBook or video every month

- Fully searchable for easy access to vital information

- Copy and paste, print, and bookmark content

Did you know that Packt offers eBook versions of every book published, with PDF and ePub files available? You can upgrade to the eBook version at packt.com and as a print book customer, you are entitled to a discount on the eBook copy. Get in touch with us at customercare@packtpub.com for more details.

At www.packt.com, you can also read a collection of free technical articles, sign up for a range of free newsletters, and receive exclusive discounts and offers on Packt books and eBooks.

Other Books You May Enjoy

If you enjoyed this book, you may be interested in these other books by Packt:

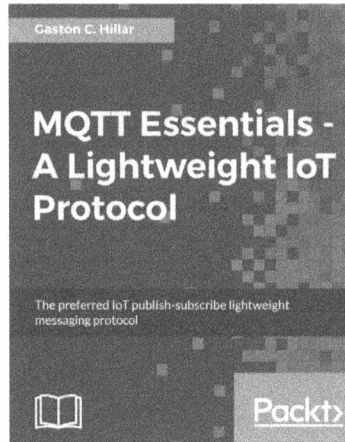

MQTT Essentials - A Lightweight IoT Protocol

Gaston C. Hillar

ISBN: 9781787287815

- Understand how MQTTv3.1 and v3.1.1 works in detail

- Install and secure a Mosquitto MQTT broker by following best practices

- Design and develop IoT solutions combined with mobile and web apps that use MQTT messages to communicate

- Explore the features included in MQTT for IoT and Machine-to-Machine communications

- Publish and receive MQTT messages with Python, Java, Swift, JavaScript, and Node.js

- Implement the security best practices while setting up the MQTT Mosquitto broker

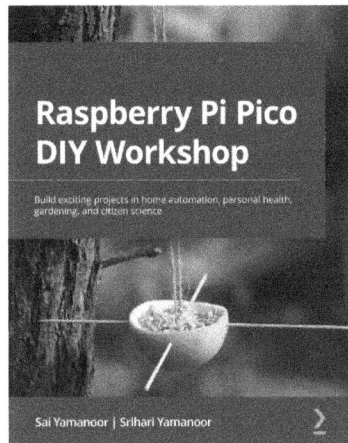

Raspberry Pi Pico DIY Workshop

Sai Yamanoor | Srihari Yamanoo

ISBN: 9781801814812

- Understand the RP2040's peripherals and apply them in the real world
- Find out about the programming languages that can be used to program the RP2040
- Delve into the applications of serial interfaces available on the Pico
- Discover add-on hardware available for the RP2040
- Explore different development board variants for the Raspberry Pi Pico
- Discover tips and tricks for seamless product development with the Pico

Packt is searching for authors like you

If you're interested in becoming an author for Packt, please visit `authors.packtpub.com` and apply today. We have worked with thousands of developers and tech professionals, just like you, to help them share their insight with the global tech community. You can make a general application, apply for a specific hot topic that we are recruiting an author for, or submit your own idea.

Share Your Thoughts

Now you've finished *Raspberry Pi and MQTT Essentials*, we'd love to hear your thoughts! Scan the QR code below to go straight to the Amazon review page for this book and share your feedback or leave a review on the site that you purchased it from.

`https://packt.link/r/1803244488`

Your review is important to us and the tech community and will help us make sure we're delivering excellent quality content.

www.ingramcontent.com/pod-product-compliance
Lightning Source LLC
Chambersburg PA
CBHW080522220326
41599CB00032B/6173